DARWIN'S PHILOSOPHY OF EMOTION

Charles Darwin is known as a biologist, geologist, and naturalist, but he was also a philosopher. This book uncovers Darwin's forgotten philosophical theory of emotion, which combines earlier associationist theories with his theory of evolution. The British associationists of the eighteenth and nineteenth centuries argued that the mind operates primarily through the association of ideas, and that emotions are strings of thoughts, feelings, and outward expressions, connected by habit and association. Charles Darwin's early notebooks on emotion reveal a keen interest in associationist philosophy. This book shows that one of his final works, *The Expression of the Emotions in Man and Animals* (1872), is a work of associationist philosophy, and analyzes Darwin's revolutionary idea: that if the associations that produce emotions can be inherited, then the theory of evolution can explain how emotions first occurred in simpler organisms and then developed and were compounded into the complex experiences humans have today.

TRIP GLAZER is an associate professor and Chair of the Philosophy Department at the University of Dayton. He has published a number of articles in journals including *Biology and Philosophy*, *Inquiry*, *Hypatia*, *Synthese*, and *Consciousness and Cognition*.

DARWIN'S PHILOSOPHY OF EMOTION

TRIP GLAZER
University of Dayton

CAMBRIDGE
UNIVERSITY PRESS

CAMBRIDGE UNIVERSITY PRESS

Shaftesbury Road, Cambridge CB2 8EA, United Kingdom

One Liberty Plaza, 20th Floor, New York, NY 10006, USA

477 Williamstown Road, Port Melbourne, VIC 3207, Australia

314–321, 3rd Floor, Plot 3, Splendor Forum, Jasola District Centre, New Delhi – 110025, India

103 Penang Road, #05-06/07, Visioncrest Commercial, Singapore 238467

Cambridge University Press is part of Cambridge University Press & Assessment, a department of the University of Cambridge.

We share the University's mission to contribute to society through the pursuit of education, learning and research at the highest international levels of excellence.

www.cambridge.org
Information on this title: www.cambridge.org/9781009538244

DOI: 10.1017/9781009538268

When citing this work, please include a reference to the DOI 10.1017/9781009538268

First published 2026

Cover image: Orangutan (Pongo pygmaeus) covering face with hand - stock photo/Ryan McVay/Getty Images

A catalogue record for this publication is available from the British Library

Library of Congress Cataloging-in-Publication Data
NAMES: Glazer, Trip author
TITLE: Darwin's philosophy of emotion / Trip Glazer, University of Dayton, Ohio.
DESCRIPTION: Cambridge, United Kingdom ; New York, NY, USA : Cambridge University Press, 2025. | Includes bibliographical references and index.
IDENTIFIERS: LCCN 2025034413 (print) | LCCN 2025034414 (ebook) | ISBN 9781009538244 hardback | ISBN 9781009538251 paperback | ISBN 9781009538268 epub
SUBJECTS: LCSH: Emotions (Philosophy) | Darwin, Charles, 1809–1882
CLASSIFICATION: LCC B105.E46 G53 2025 (print) | LCC B105.E46 (ebook)
LC record available at https://lccn.loc.gov/2025034413
LC ebook record available at https://lccn.loc.gov/2025034414

ISBN 978-1-009-53824-4 Hardback
ISBN 978-1-009-53825-1 Paperback

To Ruben

Contents

Figures

Tables

Acknowledgments

I spent nearly as long writing this book as Charles Darwin spent circumnavigating the globe on the HMS *Beagle*. My journey may not have been as daring or as significant as Darwin's, but mine was also a journey of exploration and discovery. And like Darwin's, mine would not have been possible without the help of many generous people.

Paul Morrow read multiple drafts of every chapter, providing constant encouragement and insightful feedback at every stage. Daniel Vandersommers read several chapter drafts, encouraging me to consider broader historical perspectives on many points. Andrea Scarantino provided detailed and challenging feedback on Chapters 4 and 5. Kathy Glazer caught many of my typos and helped me to make my writing more accessible.

I received helpful feedback from audiences at several conferences and workshops, including the Annual Meeting of the Ohio Philosophical Society in Akron, Ohio, in 2021; the Biennial Meeting of the International Society for Research on Emotion (ISRE) in Los Angeles, California, in 2022; the Annual Meeting of the Southern Society for Philosophy and Psychology (SSPP) in Louisville, Kentucky, in 2023; the Philosophy Department Colloquium Series at the University of Dayton in Dayton, Ohio, in 2023; the Philosophy Department Colloquium Series at the University of Arkansas in Fayetteville, Arkansas, in 2024; and finally, an online meeting of the Association Association, a reading group dedicated to the history and philosophy of associationism, also in 2024. I am thankful to all who attended these events and provided me with feedback and encouragement.

I would like to thank the University of Illinois Press for allowing me to include material that was previously published in "'Trains of thought long associated with action': Charles Darwin's Theory of Emotion," *History of Philosophy Quarterly*, 41:1 (2024), 1–22. Material from this article appears in Chapters 1 and 3 of this book.

I would also like to thank the University of Dayton for supporting my research. I was awarded two Research Council Seed Grants, in the summers of 2021 and 2022, to pursue this research.

Finally, I would like to thank my family, Sarah, Ruben, and Levi, for accompanying me on my journey.

Introduction

Charles Darwin did not have a theory of emotion – of the subjective feelings we call "anger," "fear," and "sadness." However, he developed the first evolutionary theory of emotional expression – of the smiles, pouts, and frowns that express subjective feelings. He discovered that some facial expressions of emotion are universal, and he theorized that these expressions evolved long ago to help our ancestors survive and reproduce. An expression of anger, for instance, signals one's willingness to fight, which may intimidate rivals into making concessions. Darwin's prescient theory paved the way for basic emotions theory, one of the dominant paradigms in contemporary emotions research.

This story has been told many times. Rather, this *myth* has been told many times. The story presented in the preceding paragraph is entirely incorrect. Every sentence is false.

Darwin did, in fact, have a theory of emotion. Inspired by the philosopher David Hartley, he believed that emotions are "trains" of feelings, thoughts, and behaviors, which have been joined together through past associations. Fear is the mind's simulation of the mental states and behaviors that our ancestors experienced in the face of danger. Sadness is a train of painful associations, derived from our ancestors' experience of loss, activated by our own experience of loss. According to Darwin, our emotions are lingering, transgenerational recollections of formative experiences – apparitions of the past, manifested in the present.

Darwin did believe that some facial expressions are universal, but he certainly did not "discover" this fact, given that most of his predecessors and contemporaries assumed it to be true. For example, his creationist adversaries believed that God endowed all humans (and some domesticated animals) with a universal language of emotion to facilitate communication. Even dyed-in-the-wool empiricists, who insisted that all complex mental states are learned, posited universal emotions, with universal expressions, based on universal childhood experiences. Few in Darwin's day doubted the universality of at least some emotional expressions.

Darwin was not even the first to propose an evolutionary theory of these expressions. The philosopher Herbert Spencer beat him to it. According to Spencer, emotional expressions are indeed adaptations, which helped our ancestors to survive and reproduce. But according to Darwin, many expressions serve no purpose whatsoever. Some may be adaptations (useful traits). But others are vestiges (once but no longer useful traits). And still others are spandrels (byproducts of other useful traits). Spencer emphasized the use*ful*ness of expressions; Darwin instead emphasized their use*less*ness.

Moreover, Darwin went out of his way to deny that expressions evolved as signals for communication. You can see a scowl and infer that the person is angry and willing to fight, but the person isn't scowling for the sake of showing you that they are angry or willing to fight. Darwin's expressions are "cues" (things that just happen to convey information) as opposed to "signals" (things that were naturally or conventionally selected to convey information).

Finally, Darwin's theory has little in common with modern basic emotions theory. If anything, Darwin's theory presages psychological constructionism, another contemporary paradigm that rejects the central tenets of basic emotions theory. For Darwin, as for psychological constructionists, the things we call "emotions" are simply the patterns that emerge from the complex interplay of simpler psychological processes. Emotion types like sadness, anger, and fear exist in our minds, but not in nature. They are useful schemas for explaining and predicting social behavior in real time, but not rigorous or reliable enough to serve as scientific categories.

The aim of this book is to set the story straight – to present Charles Darwin's fascinating but forgotten philosophy of emotion. Chapter 1 introduces Darwin's published book on emotional expression, *The Expression of the Emotions in Man and Animals* (1872). Chapter 2 delves into Darwin's philosophical influences, the associationists of the Enlightenment. Chapter 3 shows that Darwin developed the central ideas of *Expression* in dialogue with these philosophers. Chapters 4 and 5 detail Darwin's associationist theories of emotions and emotional expressions, respectively. Finally, Chapter 6 explains why Darwin's theory was forgotten – and why it should be remembered.

CHAPTER I

Darwin's Expression

1.1 The Origin of Darwin's *Expression*

In 1837, just one year after completing his voyage on the *Beagle*, Charles Darwin opened his first notebook on evolution (Notebook B) and wrote, "I think," followed by a sketch (B 36) (Figure 1.1). The sketch is supposed to be a simple phylogenetic tree. From an ancestral species, new species evolve.

Darwin worked out the first details of his theory of evolution in notebooks B, C, D, and E, written between 1837 and 1839.[1] At the same time, he dedicated two other notebooks, M and N, to applying his theory to humans, their minds, and their morals. Darwin referred to M and N as his "metaphysical" investigations.[2]

In the early 1840s, Darwin wrote a sketch (1842) and then an essay (1844) on natural selection but set this project aside to publish on geology and barnacles. He took up natural selection again in the late 1850s. For several years he toiled on a "big book" tentatively titled *Natural Selection*.[3]

However, in 1858 Darwin received a letter from a young naturalist, Alfred Russell Wallace, requesting Darwin's feedback on a manuscript. This manuscript proposed a theory of evolution remarkably similar to Darwin's.[4] In response, Darwin wrote up an "abstract" of *Natural Selection*, which he published under the title *On the Origin of Species* (1859).[5]

1 There are many scholarly treatments of Darwin's early theorizing about evolution. See, for instance, Smith (1960), Grinnell (1974), Herbert (1974; 1977), Egerton (1976), Schweber (1977), Kohn (1980), Ospat (1981), Colp (1986), and Hodge (1990).

2 Gruber & Barrett (1974) transcribed and analyzed notebooks M and N. Desmond et al. (2007, 38–39) highlight the importance of these notebooks in their concise biography of Darwin.

3 See Stauffer (1975) for a reconstruction of Darwin's work on this manuscript from 1856 to 1858.

4 Fichman (1981, 29–59) provides a clear overview of Wallace's theory and its differences from Darwin's.

5 Darwin recounts his version of the story in his *Autobiography* (1958, 121–122). Wallace recounts his version in *My Life* (1905, 361–364). Similar versions are repeated in Desmond et al. (2007, 63–65) and Fichman (1981, 54–55).

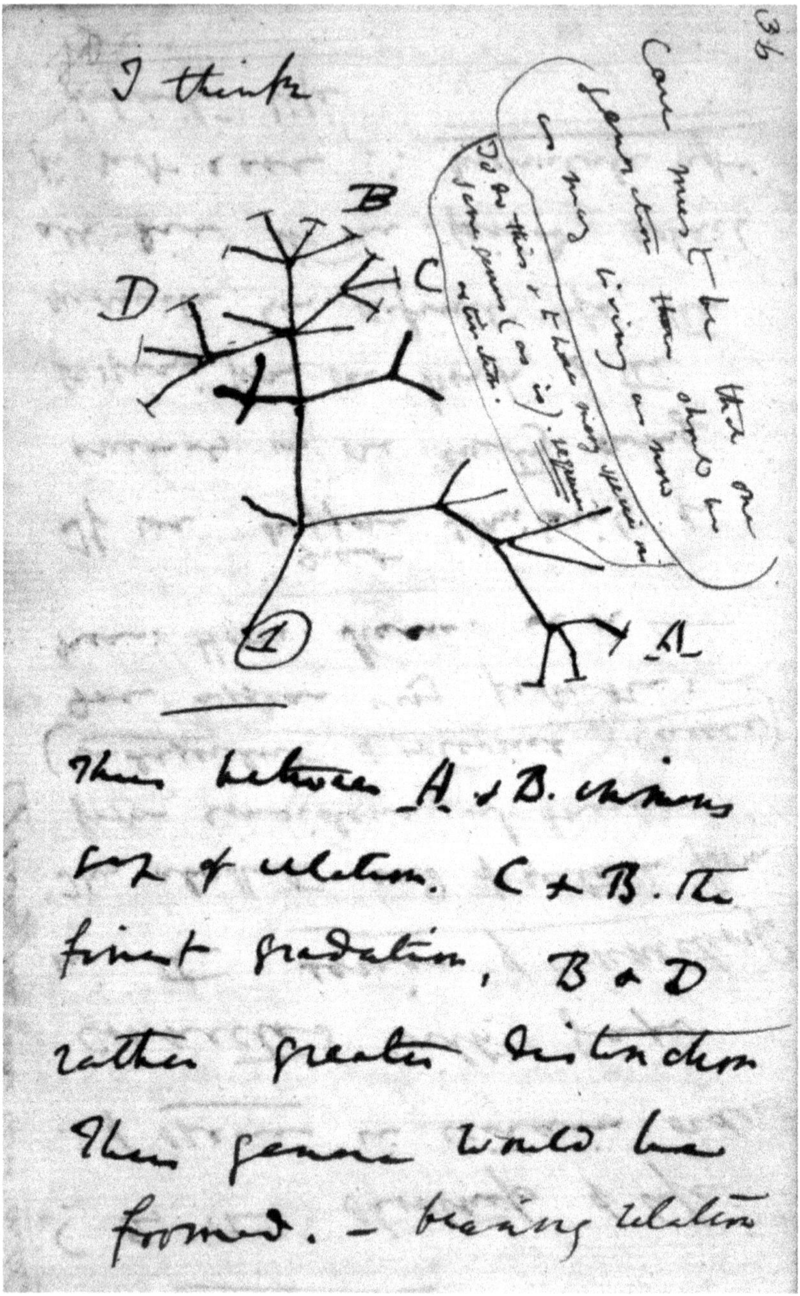

Figure 1.1 Darwin's sketch

Darwin resumed work on *Natural Selection*. The first two chapters on domestication grew so large that Darwin published them as a standalone title, *The Variation in Animals and Plants under Domestication* (1868). Darwin intended to include a chapter on humans but ultimately decided to develop it into a separate work (Darwin 1958, 131).

Three years later, Darwin published *The Descent of Man* (1871). In preparing this manuscript, Darwin consulted his old notebooks M and N, especially his summaries of philosophical debates over the moral sense. Darwin weighed in on this debate in *Descent*, arguing that human morality evolved gradually from the social instincts and emotions of mammals (Swisher 1967, 32; Richards 2009, 111–113). A planned chapter on emotional expression became too big, once again leading Darwin to develop it into a separate work.

One year later, Darwin published *The Expression of the Emotions in Man and Animals* (1872). The title is a mouthful, but it also exhibits a pleasant symmetry. "Expression" and "Man" are singular; "Emotions" and "Animals" are plural. The title combines two singular-plural pairs: "The Expression of the Emotions" followed by "in Man and Animals."[6] Like *Descent*, *Expression* contains a number of ideas that Darwin first explored in 1838–1839 and recorded in notebooks M and N.

Reflecting on Darwin's career, we can see that by end of the 1830s, Darwin envisioned a grand book on evolution, which could explain the evolution of human mind, emotion, and morality. But this single book never materialized. *Origin* (1859) sketched the theory of natural selection but said nothing of humans.[7] *Variation* (1868) developed this theory further but again said nothing of humans. *Descent* (1871) applied the theory to human mentality and morality but stopped short of emotion. *Expression* (1872), finally, took up the topic of human emotion.[8] For thirty years, Darwin's notebooks on mind, emotion, and morality collected dust in a drawer. But Darwin's dream of working through the philosophical implications of his theory of evolution were eventually realized.

6 In his *Autobiography* (1958, 131), Darwin writes the title incorrectly as *Expression of the Emotions in Men and Animals*, suggesting that this symmetry may not have been intentional.

7 Or almost nothing. The book concludes with a prediction, or perhaps a promise: "In the distant future I see open fields for far more important researches. Psychology will be based on a new foundation, that of the necessary acquirement of each mental power and capacity by gradation. Light will be thrown on the origin of man and his history" (Darwin 1859, 487).

8 To announce the publication of *Expression*, the November 16, 1872, issue of *Fun* magazine included a cartoon of Darwin (with a monkey's body) taking a woman's pulse. The woman replies, "Really, Mr. Darwin, say what you like about man, but I wish you would leave my emotions alone" (quoted in Browne 2002, 367).

1.2 The Aim of Darwin's *Expression*

The aim of *Expression* is, as its title suggests, to provide an evolutionary theory of emotional expression. More specifically, it is to explain why people and animals express their emotions as they do. Why do people express joy by smiling, sadness by pouting, and anger by frowning? Why don't they express joy by frowning, sadness by smiling, and anger by pouting? Why *this* way and not *that* way? There must be a reason – a reason that satisfies the principle of sufficient reason, or the principle that every fact requires a rational explanation.

In posing this question, Darwin takes on the creationist Sir Charles Bell, who suggests that there is no interesting explanation for why we express emotions as we do.[9] Bell speculates that God created humans with a face designed to reveal emotions: "in man there seems to be a special apparatus, for the purpose of enabling him to communicate with his fellow creatures, by that natural language which is read in the changes of his countenance" (Bell 1844, 121). But the assignment of expressions to emotions, much like the assignment of natural language words to ideas, is arbitrary. God could have made us express joy by pouting, but he instead made us express joy by smiling. That's why we smile when we feel joy. Nothing more needs to be said, and nothing more can be said.[10]

Darwin complains that the creationist explanation, if we can even call it that, is unsatisfying:[11]

> No doubt as long as man and all other animals are viewed as independent creations, an effectual stop is put to our natural desire to investigate as far as possible the causes of Expression. By this doctrine, anything and everything

[9] In 1867, five years before publishing *Expression*, Darwin wrote of his desire to challenge Bell's view in a letter to Alfred Russell Wallace: "The subject is, I think, more curious & more amenable to scientific treatment, than you seem willing to allow. I want anyhow to upset Sir C. Bell's view, given in his most interesting work "the anatomy of Expression" that certain muscles have been given to man solely that he may reveal to other men his feelings. I want to try & show how expressions have arisen" (Darwin 1867b).

[10] According to one reconstruction of Darwin's argument, the fact that animals clearly express emotions disproves Bell's assertion that humans are uniquely configured for emotional expression. However, Darwin himself notes that this argument would not cut especially deep: "If Sir C. Bell had been questioned about the expression of affection in the dog, he would no doubt have answered that this animal had been created with special instincts, adapting him for association with man, and that all further enquiry on the subject was superfluous" (1872, 11). Darwin's aim, rather, is to demonstrate that expressions (whether human or nonhuman) have natural causes.

[11] Darwin did not come up with this objection. As a university student, he attended a meeting of the Plinian Society, an undergraduate science club, in which William Browne presented this objection to Bell's theory (Browne 1995, 77).

> can be equally well explained; and it has proved as pernicious with respect to Expression as to every other branch of natural history. (Darwin 1872, 12)

The first order of business, then, is to reawaken the natural desire to investigate the causes of expressions, or the reason why expressions take the forms they do.[12] But how should that investigation proceed? To no one's surprise, Darwin suggests that "[h]e who admits on general grounds that the structure and habits of all animals have been gradually evolved, will look at the whole subject of Expression in a new and interesting light" (Darwin 1872, 12). If the theory of independent creation stops us from investigating the causes of expressions, then the theory of evolution will break the investigation wide open.

Notice that Darwin does not say here that his theory of expression will support his theory of evolution. Rather, he says that his theory of evolution will clear a path for his theory of expression. *If* we accept that humans and animals evolved, *then* we can answer the question of why people and animals express their emotions as they do. The goal of *Expression* (as stated here) is not to disprove creationism, but rather to tell the story of the causes of expressions, a story that creationism has kept under lock and key.

Darwin teases two examples of expressions whose causes may become known, once we entertain the evolution of species:

> With mankind some expressions, such as the bristling of the hair under the influence of extreme terror, or the uncovering of the teeth under that of furious rage, can hardly be understood, except on the belief that man once existed in a much lower and animal-like condition. (Darwin 1872, 12)

Notice, however, that our descent from a nonhuman ancestor doesn't yet explain why we bristle in terror and snarl in rage. It merely pushes the question back. Why did this ancestor express terror and rage in these ways? Why not express terror and rage in entirely different ways?

To explain why expressions take the forms they do, Darwin proposes three principles.[13]

> 1. *The principle of serviceable associated Habits.* – Certain complex actions are of direct or indirect service under certain states of the mind, in order to relieve or gratify certain sensations, desires, etc.; and whenever the same state of mind is induced, however feebly, there is a tendency through the force of habit and association for

[12] Campbell (1997a) discusses the evolving meaning of "cause" in Darwin's day.
[13] My summaries of these principles were previously published in Glazer (2024).

> the same movements to be performed, though they may not then be of the least use.[14] (Darwin 1872, 28)

Some animals puff out their fur or feathers when threatened. This action is serviceable in that it helps the animal to survive, perhaps by making it appear larger and more formidable to hungry predators and angry competitors. If the animal performs this serviceable action repeatedly, then "through the force of habit and association" it may become an automatic, involuntary response to danger. Even if the danger is a human hunter with a gun, the animal may involuntarily puff out its fur or feathers one last time before perishing. The action, once serviceable, was no longer of service in this case.

> 2. *The principle of Antithesis.* – Certain states of the mind lead to certain habitual actions, which are of service, as under our first principle. Now when a directly opposite state of mind is induced, there is a strong and involuntary tendency to the performance of movements of a directly opposite nature, although these are of no use; and such movements are in some cases highly expressive. (Darwin 1872, 28)

Dogs, when feeling aggressive, tense their muscles in preparation to fight. So when they feel the opposite of aggression, affection, their muscles tend to relax (Darwin 1872, 271). Darwin analyzes the shrug of indifference similarly. When we make up our minds to do something, we tend to extend our arms in pursuit of that end. So when we can't make up our minds, we toss our shoulders up, drawing our arms closer to our bodies, palms up (Darwin 1872, 271). The opposite feeling prompts the opposite reaction.

> 3. *The principle of actions due to the constitution of the Nervous System, independently from the first of the Will, and independently to a certain extent of Habit.* – When the sensorium is strongly excited, nervous force is generated in excess, and is transmitted in certain definite directions, depending on the connection of the nerve-cells, and partly on habit: or the supply of nerve-force may, as it appears, be interrupted. Effects are thus produced which we recognise as expressive. This third principle may, for the sake of brevity, be called that of the direct action of the nervous system. (Darwin 1872, 28–29)

An early review of *Expression* criticizes this principle for being too vague (Bain 1873, 698). In a private letter to the reviewer, Darwin concedes that the criticism is "perfectly just" (Darwin 1873). Indeed, this principle has

[14] To keep things simple for now, I have included only part of the first principle. I will have more to say about the rest of the principle later, especially in §3.8.2.

proven the most vexing for interpreters.[15] The idea seems to be that when an animal perceives something that requires a rapid response, nerve force is generated; and if more nerve force is generated than is needed for the response, then the excess spills over into unserviceable expressive behaviors. Darwin observes that rabbits and cows tend not to vocalize, except when suffering greatly. He hypothesizes that excess nerve force is released through the respiratory system, resulting in purposeless cries (1872, 83).

If you had to guess Darwin's theory of emotional expression before reading *Expression*, then you might have guessed it to be that expressions are adaptive – that they help animals to survive and reproduce. However, as many scholars have pointed out, Darwin instead emphasizes the *uselessness* of many emotional expressions.[16] Granted, some expressions began as voluntary purposive movements, but in becoming involuntary expressions, they became largely purposeless. They became vestiges. Darwin writes in *Descent* that *Origin* had overlooked "many structures which appear to be, as far as we can judge, neither beneficial nor injurious" (1871, I:152). *Expression* aims to correct *Origin*'s mistake by analyzing behaviors that do not serve any survival function.[17]

So why do we express our emotions as we do? Darwin's three principles enable us to identify the "cause" of every expression – the rational explanation for why the expression is *this* and not *that*. The creationists supposed that no rational explanation is possible, yet Darwin has shown that one is. Darwin concludes *Expression* with a modest plea for further research on the subject:

> To understand, as far as is possible, the source or origin of the various expressions which may be hourly seen on the faces of the men around us, not to mention our domesticated animals, ought to possess much interest for us. From these several causes, we may conclude that the philosophy of our subject has well deserved the attention which it has already received from several excellent observers, and that it deserves still further attention, especially from any able physiologist. (1872, 367)[18]

[15] Gruber writes, "The third principle is something of a vague catch-all" (Gruber & Barrett 1974, 326). Green notes that "Darwin says little to elucidate this principle" (2007, 113).

[16] See Burkhardt (1985, 359) and Richards (2009, 114) for historians' perspectives, Fridlund (1992) for a psychologist's perspective, and Campbell (1997a, 454) for a philosopher's perspective.

[17] Burkhardt (1985, 357) made the classic case for this interpretation. Vestigial traits – those that once but no longer serve a function – offer better evidence for evolution than do adaptive traits, since creationism leaves no room for them.

[18] Compare Darwin's plea to John Stuart Mill's: "It is conceivable that a scientific theory of Fear may one day be constructed, but it must evidently be the work of physiologists, not of metaphysicians" (Mill 1978, XI:362). I have not been able to confirm that Darwin read this passage, which comes from Mill's review of Alexander Bain's philosophy of mind and emotion, but it is possible, given the

1.3 The Structure of Darwin's *Expression*

Filling not quite 400 pages, *Expression* is shorter than *Origin* and *Descent*, but it can hardly be called short. The book is divided into an introduction and fourteen chapters, the last of which doubles as a conclusion.

The introduction begins with a literature review. Darwin summarizes prior work on emotional expression and situates his contributions within the literature.[19] He starts by dismissing the physiognomic tradition as all but entirely useless (1872, 1). Physiognomy, not to be confused with physiology, is the debunked pseudo-science of inferring character and personality traits from the shape and structure of the face.[20] Darwin then credits Bell and other physiologists with describing many expressions correctly, but then critiques them for failing to explain why these expressions take the forms they do. "[Bell] does not try to explain why different muscles are brought into action under different emotions; why, for instance, the inner ends of the eyebrows are raised, and the corners of the mouth depressed, by a person suffering from grief or anxiety" (1872, 3). Finally, Darwin credits two philosophers, Alexander Bain and Herbert Spencer, with getting many explanations right. He registers agreement with Bain's principles yet notes that they are "too general to throw much light on special expressions" (1872, 9). He says of Spencer's explanations that they are "of the highest importance in throwing light on our subject" (1872, 11), yet "the chief interest and difficulty of the subject lies in following out the wonderfully complex results" (1872, 9). Neither Bain nor Spencer captured the complexity and nuance of expression.

There is, in Darwin's literature review, a kind of Hegelian synthesis. The physiognomic tradition contributed nothing. The physiological tradition contributed descriptions of expressions, but not explanations. The philosophical tradition contributed explanations, but not descriptions. Darwin's *Expression*, finally, provides detailed descriptions alongside robust explanations.

similarities of these passages and Darwin's interest in Bain's philosophy in the late 1860s, combined with his professed inability to parse Bain's prose.

19 Interestingly, Darwin omits nearly all the sources discussed in his 1838–1839 notebooks on emotional expression, despite the fact that he consulted these notes and drew liberally from them in preparing *Expression*. I have more to say about these omitted sources in §§3.2–4. See Manier (1978) for a careful study of Darwin's notebook sources.

20 Darwin's half-cousin Sir Francis Galton dabbled in physiognomy. See Shapin (1975) for a critical reflection on nineteenth-century debates over phrenology.

Darwin would prefer for his contribution to be cast in Baconian terms.[21] He follows the "true but unattempted" method of "ascending continually and gradually" from detailed descriptions of specific expressions to "the most general axioms" of expression (Bacon 1901, XIX). Indeed, he asserts, somewhat disingenuously, that "I arrived ... at these three Principles only at the close of my observations" (1872, 27).[22]

In his introduction, Darwin also reports the results of an informal experiment. He sent a questionnaire to recipients around the world, asking how the locals express their emotions. The questionnaire is quaint, and worth reproducing in part (the complete questionnaire includes sixteen or seventeen questions[23]):

> 1. Is astonishment expressed by the eyes and mouth being opened wide, and by the eyebrows being raised?
> 2. Does shame excite a blush when the colour of the skin allows it to be visible? and especially how low down the body does the blush extend?
> 3. When a man is indignant or defiant does he frown, hold his body and head erect, square his shoulders and clench his fists?
> 4. When considering deeply on any subject, or trying to understand any puzzle, does he frown, or wrinkle the skin beneath the lower eyelids? (Darwin 1872, 15–16)

The results of the survey confirmed, to Darwin's satisfaction, that many emotional expressions are universal, occurring in all human societies.[24]

It would be a mistake to say that Darwin "discovered" the universality of emotional expression, however, since belief in the universality of expression was already widespread. As a young man, Darwin studied the philosophy of Dugald Stewart, who claimed that languages evolved from innate, universal emotional expressions (Manier 1978, 38). Bell and other creationists likewise claimed that expressions are universal – a natural language that enables communication across language barriers. Despite widespread belief in the universality of expressions, Darwin suggests that it

[21] See Porter (2018a) for an insightful analysis of Bacon's influence on Erasmus Darwin and the general culture of science into which Charles Darwin was born.

[22] Darwin claims that his observations began in 1838 and continued for many years. Yet Darwin's notebook M, completed in 1838, includes formulations of his first two principles.

[23] See Freeman & Gautrey (1972; 1975) for critical examinations of the different versions of the query that exist. Although Darwin lists sixteen questions in *Expression*, he sent queries with seventeen questions to many of his correspondents. *Expression*'s list leaves out the only question about vocal expressions of emotion.

[24] Darwin perhaps overstates his conclusion here. The answers to his queries revealed some cultural differences in expressions, and he discusses these differences throughout *Expression*. I discuss this point further in §3.7.

is "highly important to ascertain whether the same expressions and gestures prevail, as has often been asserted without much evidence, with all the races of mankind" (Darwin 1872, 15). Thus, although Darwin didn't discover the universality of emotional expression, he deserves some credit for seeking out independent empirical justification.

Following the introduction, chapters 1–3 present Darwin's three principles, one per chapter. Chapter 1 is entitled "General Principles of Expression." Chapter 2 is "General Principles of Expression – Continued." And chapter 3 is "General Principles of Expression – Concluded." These chapters set a theoretical foundation for the rest of *Expression*.

Chapters 4 and 5 focus on the expressions of nonhuman animals. Darwin enumerates dozens of examples of nonhuman animal expressions and applies his three principles to explain why they take the forms that they do. As I've already mentioned, he explains the bristling of fur or feathers using the first principle (1872, 95), the wagging of the dog's tail using the second (1872, 119), and the wailing of otherwise silent animals using the third (1872, 83).

Chapters 6–13 turn to the expressions of humans. Chapter 6 serves as a natural transition from nonhumans to humans by highlighting expressions that are unique to the latter. Darwin in fact focuses on one, "weeping." He concludes the chapter with a summary of his view:

> To sum up this chapter, weeping is probably the result of some such chain of events as follows. Children, when wanting food or suffering in any way, cry out loudly, like the young of most other animals, partly as a call to their parents for aid, and partly from any great exertion serving as a relief. Prolonged screaming inevitably leads to the gorging of the blood-vessels of the eye; and this will have led, at first consciously and at last habitually, to the contraction of the muscles round the eyes in order to protect them. At the same time the spasmodic pressure on the surface of the eye, and the distension of the vessels within the eye, without necessarily entailing any conscious sensation, will have affected, through reflex action, the lacrymal glands. Finally, through the three principles of nerve-force readily passing along accustomed channels – of association, which is so widely extended in its power – and of certain actions, being more under the control of the will than others – it has come to pass that suffering readily causes the secretion of tears, without being necessarily accompanied by any other action. (Darwin 1872, 176)

Darwin repeats this explanation in the concluding chapter (1872, 362) to highlight both the complexity of the causal history of some expressions and the fecundity of his three principles.

Chapters 7–13 enumerate common and uncommon expressions in humans, organized by emotion.[25] Chapter 7 reviews expressions of grief and despair. Chapter 8 covers love and joy. Chapter 10 turns to anger and rage. As in the chapters on nonhuman animals, Darwin surveys diverse expressions and explains each using one or more of his principles. For instance, Darwin explains the shrugging of the shoulders using the principle of antithesis (1872, 271). We habitually grasp when determined. Therefore, we may involuntarily do the opposite of grasping (shrugging) when feeling the opposite of determination (powerlessness or indifference).

Chapter 14, finally, is entitled "Concluding Remarks and Summary." Darwin begins this chapter by reiterating his general strategy: He surveys examples of emotional expressions and attempts to account for each using one or more of his general principles of expression. These principles lay bare the "cause" of the expression, or the reason why the expression takes that form as opposed to any other.

Chapter 14 also includes one of the most famous passages from *Expression*:

> The free expression by outward signs of an emotion intensifies it. On the other hand, the repression, as far as this is possible, of all outward signs softens our emotions. He who gives way to violent gestures will increase his rage; he who does not control the signs of fear will experience fear in a greater degree; and he who remains passive when overwhelmed with grief loses his best chance of recovering elasticity of mind. (Darwin 1872, 366)

This passage presents what is sometimes called the facial feedback hypothesis, or the idea that the expression (or repression) of an emotion can intensify (or weaken) the experience of that emotion. Screaming when angry increases the temperature, so to speak, while resisting the urge to scream is calming. In a classic experiment (Strack et al. 1988), subjects were presented with a series of cartoons and asked to rate the humor of each while holding a pen either between their teeth (inducing a smile) or between their lips (inducing a pout). The study found that the "smiling" subjects produced higher ratings than the "pouting" subjects, supporting the hypothesis. Although many credit Darwin with proposing the facial feedback hypothesis, Darwin credits its discovery to Louis Pierre Gratiolet, the French neurologist who first named the five lobes of the brain.[26] Yet, as

[25] Darwin's organization of emotions is curious. However, I argue in §4.4 that Darwin's list and organization of emotions may have been influenced by the philosopher Thomas Brown.

[26] The claim that Darwin proposed it is a textbook example of the Matthew effect, or the tendency to distribute credit according to esteem, with the more famous enjoying more credit and the less famous enjoying less credit than deserved (Merton 1968). The name comes from the parable of the talents from the gospel of Matthew: "For whoever has will be given more, and they will have an

we shall see later, Darwin's grandfather Erasmus Darwin also defended this hypothesis, as did many other Enlightenment philosophers.

1.4 The Images of Darwin's *Expression*

Flipping through the pages of *Expression*, one's eyes are drawn to the many striking, and at times chilling, images contained within it. Twenty-one of these images are illustrations. Thirty are photographs.

The first three illustrations are anatomical diagrams, reproduced from other books on expression. These illustrations label the muscles of the human face. The first illustration is taken from one of Bell's works. (To Bell's claim that God created these muscles as instruments of expression, Darwin replies that these muscles serve other functions, besides expression.) The second and third are taken from Friedrich Gustav Jakob Henle, the early advocate of germ theory.

The next fifteen illustrations are engravings of animals: a cat expressing affection, a dog expressing anger, an ape expressing disappointment, and so on. Darwin commissioned several artists to create these engravings and was heavily involved in the creative process. For instance, Darwin asked artist Joseph Wolf to rework the ears of an ape, and, still unsatisfied with the result, asked the engraver to rework them further (Prodger 2009, 143). Apparently, Wolf reached his limit working with Darwin, and Darwin had to commission another artist, Thomas William Wood, to bring Wolf's illustrations of a dog and a cat in line with Darwin's expectations (Prodger 2009, 147–149). In another instance, Darwin commissioned artist Briton Riviere to illustrate a dog showing affection. When Riviere refused to incorporate some of Darwin's creative input, Darwin asked Wood to rework the illustration. This time Wood refused. Ultimately, Darwin and Riviere found a compromise (Prodger 2009, 152–153) (Figure 1.2). Phillip Prodger suggests that these kerfuffles between Darwin and his artists reflect a deeper question: Is the artist's eye or the scientist's understanding the better guide to optical truth (2009, 153)?

The final three illustrations are engravings of human expressions, based not on artists' drawings but on photographs. For the sake of exposition, I will discuss the photographs included within *Expression* first, and then return to these engravings afterward.

abundance. Whoever does not have, even what they have will be taken from them" (Matthew 25:29 NIV).

Figure 1.2 Engraving of a dog

In addition to the engravings, *Expression* features thirty photographs. *Expression* is Darwin's only book to include photographs, and it is among the first books in history to print photographs directly to the page (Prodger 2009, 41, 108). Previously, publishers had to produce photographs separately and then glue them into the book, an expensive and time-consuming process. To make the inclusion of photographs feasible, Darwin suggested a recently engineered method, called heliotypy (Prodger 2009, 109). Basically, a photographic negative is exposed onto a gelatin film, which is then hardened using chemicals. The resulting plate can be used to mass produce the photograph. However, owing to uncertainty about the cost and effectiveness of this new process, Darwin also commissioned engravings of three photographs, mentioned previously, which were included in *Expression* as engravings rather than as photographs.

Expression's thirty photographs can be divided thematically into three groups. Nearly half (fourteen) of the photographs are of babies or young children. Of these, ten were taken by the photographer Oscar Rejlander, famous for his photomontages. Darwin wanted to include photographs of children because he believed their expressions to be entirely natural, free of artifice (Prodger 2009, 105; Lorch & Hellal 2010). The most famous of the child photographs, unofficially titled "Ginx's Baby," depicts a baby crying in distress (Figure 1.3). Interestingly, the photograph included in

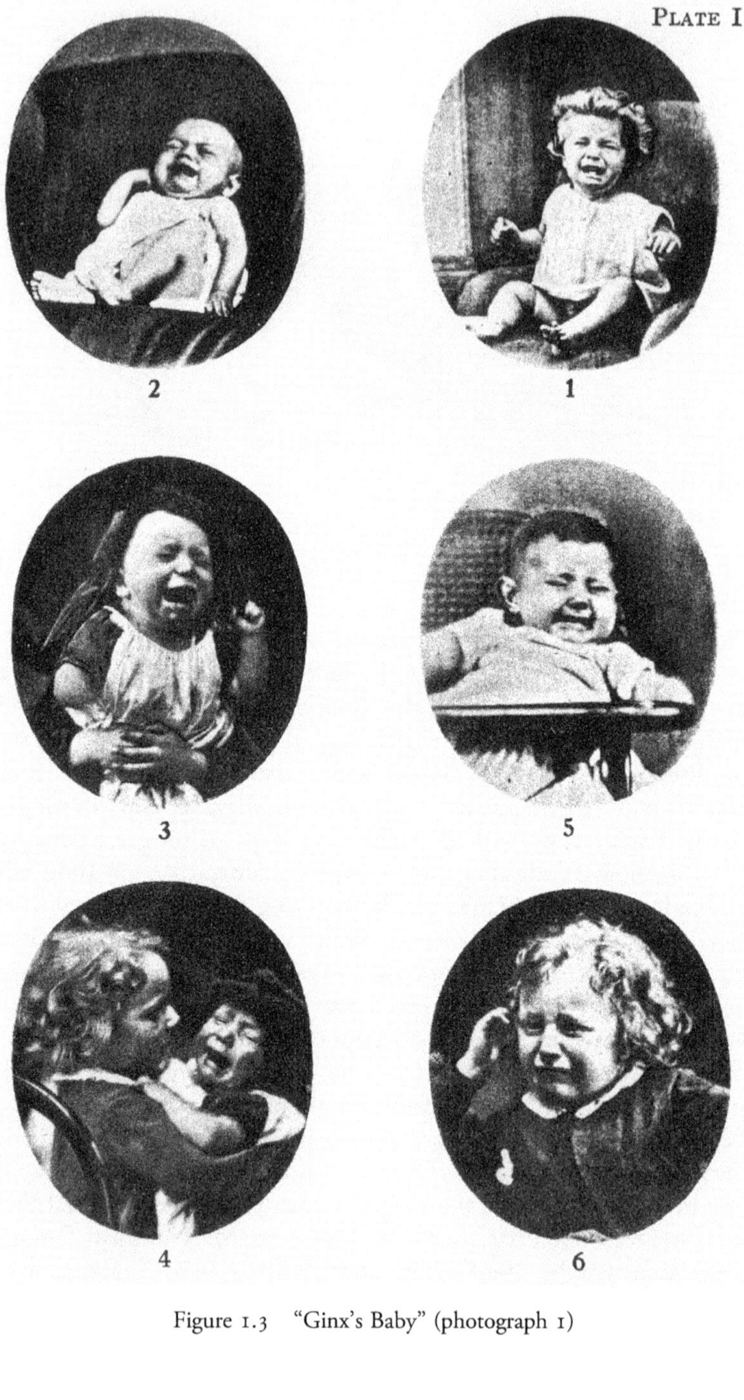

Figure 1.3 "Ginx's Baby" (photograph 1)

the first edition of *Expression* was not the original photograph. Rather, it was a photograph of a chalk drawing of the original photograph (Prodger 2009, 107–108). Some subsequent editions included the original photograph, but many others include the photograph of the drawing of the photograph.

In the second group of photographs, the photographer Rejlander steps in front of the camera with his wife Mary and another young man to perform dramatized expressions (Browne 2002, 367). These expressions are admittedly hokey, but as Prodger observes, Rejlander's balding head provides a clear view of the wrinkles of the forehead, which Darwin takes care to emphasize in his descriptions of expressions (Prodger 2009, 195).

The third set of photographs is by far the most famous. These are photographs of people living with mental illness first published by the French neurologist Guillaume-Benjamin Duchenne de Boulogne. Duchenne artificially induced facial expressions in his subjects by galvanizing the muscles in their face with an electric current. This method enabled Duchenne to determine which muscles contracted in various facial expressions of emotion. Duchenne published eighty-four photographs in his *Mechanism of Human Facial Expression* (1862), a book that influenced Darwin's work on emotional expression. When Darwin wrote to Duchenne for permission to reprint several of these photographs, Duchenne granted permission and declined payment (Prodger 2009, 83). Darwin included four of Duchenne's photographs and engravings of two others.

Owing to misgivings about the heliotype process, Darwin commissioned engravings of three photographs, and included these engravings in place of the original photographs (Figures 1.4 and 1.5). Besides the two engravings of Duchenne's "old toothless man," there is an engraving of a photograph of a woman with wild hair. Prodger points out that, as with the engravings of animals, Darwin directed his engraver to make considerable alterations to the original photographs (Prodger 2009, 97–98). The photograph of the woman with wild hair shows few wrinkles around the forehead, yet the engraving shows pronounced wrinkles. Darwin instructed his engraver to remove the electrodes from Duchenne's photographs of the "old toothless man" and to add definition to the wrinkles around the forehead and neck (Prodger 2009, 87). The resulting images are less chilling than the original photographs, but no less striking.

The historians of science Lorraine Daston and Peter Galison (2007) argue that photographic technology ushered in a new conception of objectivity in science. According to the norm of "mechanical objectivity,"

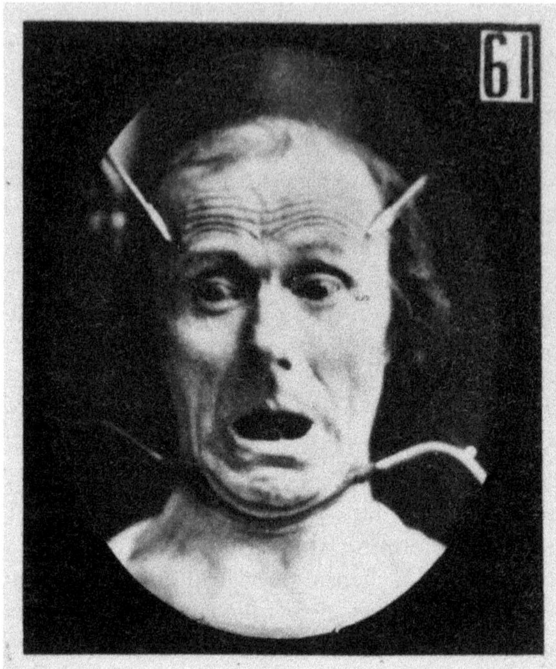

Figure 1.4 Terror (photograph)

Figure 1.5 Terror (engraving)

the scientist must refrain from inserting their subjectivity into the image-making process. They must allow an impartial machine to record nature in its pure, unadulterated form. Darwin adheres to an older conception of objectivity, thinking that he must use his understanding and insight to improve upon faulty or misleading images.[27]

1.5 The Puzzle of Darwin's *Expression*

Darwin's *Expression* has puzzled scholars for 150 years. What puzzles them is that Darwin got several things wrong, which, frankly, he should have gotten right, given his theory of evolution by natural selection. These mistakes are, to use a sports analogy, unforced errors – mistakes that Darwin could have and should have avoided, given his knowledge at the time. In what follows, I present three unforced errors that have puzzled scholars since the publication of *Expression*:

1. Darwin was wrong to focus on habit.
2. Darwin was wrong to sidestep emotion.
3. Darwin was wrong to downplay communication.

1.5.1 Habit

The first puzzle is Darwin's focus on habit. Recall that all three principles of expression refer to habit. The first principle states that voluntary actions become involuntary through force of habit. The second states that, given a habit of expressing an emotion one way, the opposite emotion will be expressed in the opposite way. And the third states that the flow of excess nerve energy depends "partly on habit" (Darwin 1872, 29). Darwin's focus on habit is surprising, however, given his discovery of natural selection.

Darwin was not the first to propose that species evolve. His grandfather, Erasmus Darwin, famously considered the evolution of species in his *Zoonomia* (1796):

> Would it be too bold to imagine, that in the great length of time, since the earth began to exist, perhaps millions of ages before the commencement of the history of mankind, would it be too bold to imagine, that all warm-blooded animals have arisen from one living filament, which THE GREAT FIRST CAUSE endued with animality, with the power of acquiring new

[27] See Porter (2018b) for an analysis of Erasmus Darwin's epistemic uses of images, which may have influenced Charles.

> parts, attended with new propensities, directed by irritations, sensations, volitions, and associations; and thus possessing the faculty of continuing to improve by its own inherent activity, and of delivering down those improvements by generation to its posterity, world without end! (1796, 39.4.8)

Erasmus was right *that* species evolve. But he was wrong about *how* species evolve. He proposed that evolution occurs through the inheritance of acquired characteristics, including habits (1796, 39.4.8).[28] If food on the savanna becomes scarce, then giraffes may have to strain their necks to reach the leaves on higher branches. The result of this frequent straining may be elongated necks. And the result of elongated necks may be offspring with slightly longer necks – that is, the inheritance of an acquired characteristic. With each generation of straining, giraffes' necks get longer and longer.

A similar theory of evolution was proposed a decade later by Jean-Baptiste Lamarck (1809), who coined the term "biology." Today the theory that evolution occurs through the inheritance of acquired characteristics is known as "Lamarckism."[29]

Charles Darwin corrected his grandfather's mistake. He discovered the true mechanism of evolution, natural selection. Darwin made this discovery in 1838–1839 but waited to publish it until he learned that Alfred Russell Wallace had independently discovered natural selection. The full title of Darwin's 1859 masterpiece is *On the Origin of Species by Means of Natural Selection, or The Preservation of Favoured Races in the Struggle for Life*.[30] The title's gloss on natural selection summarizes the mechanism nicely. If there is variation among the members of a species, and if there is a struggle for life (against predators, starvation, the elements, etc.), then the organisms with advantageous traits are more likely to survive and reproduce than those with disadvantageous traits. A giraffe born with a slightly longer neck than his siblings will be able to eat the leaves that his sibling can't reach, meaning that he is less likely to starve. Assuming, finally, that these advantageous traits can be passed down to offspring, it follows that with each subsequent generation, more and more offspring will be born with the advantageous traits. The giraffes with the longest necks are the

28 See Zirkle's (1935, 418) classic treatment.

29 Osborn (1894, 153–155) argues that Lamarck was ignorant of Erasmus Darwin's theory and came to the same ideas independently. See Burkhardt (1977) for a detailed study of Lamarck's theories and contributions to biology.

30 It was Darwin's publisher, John Murray, who came up with the title. Darwin wanted to call it *An Abstract of an Essay on the Origin of Species and Varieties through Natural Selection* (Browne 2002, 81). Thank goodness for good editors.

most likely to become parents, and of their offspring, those with still longer necks are the most likely to become parents, and so on. This is evolution by natural selection. Variation plus struggle plus inheritance equals evolution.

A pithy way of putting the difference between these theories is that for Erasmus and Lamarck, adaptation occurs within an organism's lifetime (via learning), whereas for Charles, adaptation occurs between generations (via differential reproduction).

What puzzles scholars about *Expression*, then, is that by analyzing expressions as acquired habits, Charles Darwin seems to eschew his own correct theory of evolution and to embrace Erasmus's and Lamarck's incorrect theory. Some scholars dismiss *Expression* as an unfortunate step backward in Charles's thinking about evolution.[31]

But for the record, Charles agreed with Erasmus and Lamarck that acquired traits can be inherited. In *Origin* (1859, 134), Darwin claims that the continued use of organs "strengthens and enlarges" them, that "disuse diminishes them; and that such modifications are inherited." We find this claim again in *Variation* (1868a, II:293–318), *Descent* (1871, I:152–153), and of course *Expression* (1872, 39).[32]

But the use and disuse of organs is only one source of heritable variation. Charles also acknowledges the "spontaneous appearance" of variation (e.g., 1868a, II:314).[33] In any given litter of giraffe calves, some will have longer necks than others. The longer-necked calves will outreproduce the shorter-necked calves. Even in *Origin*, he argued that the honeybee's hive-building instincts "can be explained by natural selection having taken advantage of numerous, successive, slight modifications of simpler instincts" (1859, 235).[34] By the time of the modern synthesis of

[31] See, for instance, Browne (2002, 369) and Barrett (2015, 67).

[32] Richards (1992, 94), Smith (2010), and Hernández-Avilez & Ruiz-Gutiérrez (2023) analyze Erasmus's influence on Charles's theory of evolution. Winther (2000) provides a careful analysis of Charles's evolving views of inheritance and variation.

[33] Swisher (1967, 34–35) argues that Darwin's early writings on evolution placed equal emphasis on inherited habit and chance variation, but that by 1859 he placed greater emphasis on the latter. Winther (2000) agrees, highlighting a passage from *Variation* in which Darwin writes that "it is far more probable that the webbing [between the toes of water dogs], like the membrane between the toes of certain pigeons, spontaneously appeared and was afterwards increased by the best swimmers and the best snow-travelers being preserved during many generations" (Darwin 1868a, II:500–501; quoted in Winther 2000, 439).

[34] Darwin's erroneous theory of the biology of inheritance, pangenesis, allows for both innate and acquired traits to be inherited (1868a, II:357–404). The idea, briefly, is that each part of an organism emits particles, called gemmules, which encode information about itself. A giraffe with a long neck, whether it got that long neck spontaneously or from frequent stretching, has long-neck gemmules that collect in the gametes that unite with other gametes in sexual reproduction to form

Darwin's theory of natural selection and Mendelian genetics, natural selection was understood to operate on those "spontaneous variations" that result from genetic differences, rather than those variations that are acquired in the course of life.

Still, it is curious that Charles insists that many emotional expressions originated as acquired heritable habits, when he could just as easily have proposed that these expressions originated as spontaneous modifications of instincts. Suppose that some mice instinctually express fear by running away from danger and that other mice instinctually express fear by running toward danger. Obviously, those that run away are more likely to survive and reproduce. The mice that run toward danger will die before reproducing. The advantageous instinct – running away from danger – is more likely to be passed down than the disadvantageous instinct – running toward danger – resulting in future generations of mice that instinctually run away from danger when afraid.

In the twentieth century, the Darwinian ethologists Konrad Lorenz, Nikolaas Tinbergen, and Karl von Frisch would win a Nobel Prize for their analysis of instincts as "fixed action patterns," or short sequences of actions that unfold automatically and invariably in response to specific stimuli.[35] These patterns arise not from habit but from accumulations of slight modifications of innate instincts. Darwin could have given this kind of analysis in *Expression*, but instead he appealed to the notion of acquired heritable habits.

To review: Charles Darwin analyzes expressions as acquired habits that can be passed down. But he could have analyzed expressions as innate instincts that can be passed down. His theory of natural selection gives a plausible explanation for why people and animals instinctually express their emotions as they do. Our expressions helped our ancestors to survive; alternative expressions helped our ancestors' kin to perish. Why, then, does Darwin insist on habit? That is the first puzzling mistake of *Expression*.

1.5.2 Emotion

The second puzzling mistake of *Expression* is that it sidesteps emotion. Darwin discusses emotional expressions in great detail, yet he says

zygotes. Offspring develop according to the blueprint provided by their parents' gemmules. Thus, long-necked parents give birth to long-necked children, who may have longer necks, again spontaneously or by stretching, and so have longer-necked grandchildren.

35 See the Nobel Prize in Physiology or Medicine 1973 press release at www.NobelPrize.org. See also Burkhardt (2005) for a scholarly account.

surprisingly little about the emotions themselves. An entire chapter is dedicated to explaining why we weep, for instance, yet not even a page is dedicated to explaining why we feel sadness in the first place.

Darwin's reticence about emotion has led many scholars to conclude that Darwin had no theory of emotion. The historian of emotion Thomas Dixon asserts exactly that:

> [Darwin] did not have any theory of emotions. He took the term "emotions" to stand for a range of mental states such as fear, anger, pride, shame and the like, but he never tried to explain the origin or functions of emotional feelings, nor did he try to define or classify the emotions per se. (2003, 166)

The philosopher of emotion John Deigh concurs: "Darwin himself was unconcerned with the question of the nature of emotions.. . . Darwin was chiefly concerned with involuntary expressions of emotion" (2004, 14–15). And the psychologists of emotion Maria Gendron and Lisa Feldman Barrett are of the same mind: "Darwin never intended to craft a model of emotion" (2009, 322). Why did Darwin write *The Expression of the Emotions in Man and Animals* as opposed to *The Emotions in Man and Animals*?

The problem is not simply that Darwin didn't explain emotions. Nature is complicated. Minds are complicated. We can't expect Darwin to explain everything. The problem runs deeper. The problem is that Darwin's theory *can't* explain emotions – that Darwin's account of the evolutionary origins of emotional expressions leaves no room to explain the evolutionary origins of emotional feelings.

To get a sense for this worry, it will be helpful to give a rough sketch of contemporary Darwinian theories of emotion. These theories claim that emotions evolved (at least in part) because their expressions are adaptive. Animals experience fear, for instance, because the expression of fear, fleeing, helped the animals' ancestors to survive. On this view, emotions are "affect programs," or dispositions to respond to life's common challenges in beneficial ways, including those behaviors that we call "expressions."[36] The *feeling* of the emotion is, perhaps, the sensation of the body

[36] Philosophical formulations include Griffiths (1997), Charland (2002), Prinz (2004), Green (2007), Scarantino & Griffiths (2011), Scarantino (2015), Glazer (2018a), and Hutto, Robertson, & Kirchhoff (2018). Scientific formulations include Plutchik (1962), Tomkins (1962; 1963), Ekman (1992), Panksepp (1992), and Keltner et al. (2019). I discuss these theories further in §4.5.

mobilizing its resources in preparation to react.[37] In any case, the usefulness of the emotion derives largely from the usefulness of its expression. We experience emotions (at least in part) for the sake of their expressions.

Notice how different contemporary Darwinian theories are from Darwin's original theory. In *Expression*, Darwin denies that expressions are always, or even often, adaptations. He claims that these behaviors first occurred for reasons that have nothing to do with emotions, later became associated with emotions, and now occur involuntarily and without purpose. Emotions, then, are little more than conditioned stimuli. Recall Pavlov's dog. When he sees food, he salivates. If he hears a bell every time his food is served, then he will salivate at the sound of a bell, even if no food is served. Darwin's dog is similar. When he sees an enemy, he growls. If he also feels angry every time he sees an enemy, then he will growl upon feeling angry, even if no enemy is present. But what's the point of having the emotion in the first place? It is not experienced for the sake of the expression. It does nothing in Darwin's theory, other than to serve as a mysterious middleman.

The philosopher John Dewey was perhaps the first to identify this puzzle in his critical examination of Darwin's theory:

> I wish to point out that in the case of "serviceable associated habits," the principle of explanation actually used, whatever the form of words employed, is that of survival, . . . of acts originally useful not qua expressing emotion, but *qua* acts – as serving life.. . . *The reference to emotion in explaining the attitude is wholly irrelevant; the attitude of emotion is explained positively by reference to useful movements.* (Dewey 1894, 555–556)

Dewey's language is a little confusing. In contemporary philosophical terms, an "attitude" is a mental state, a thought. But for Dewey, an "attitude" is an action or action pattern, in this case the so-called expression. The italicized portion of the text asserts that emotions don't figure into the explanation of expressions. Dogs growl because their ancestors growled to fend off enemies. Anger just happened to be at the right place at the right time.

Dewey then suggests that emotions can be cut from Darwin's theory entirely. He writes:

> In the discussion of movements in animals . . . the reference to emotion is not even nominal. It is a matter of "satisfaction of desire" and "relieving

[37] This is William James's (1884) classic theory, which I discuss in §6.3. Shargel & Prinz (2017) defend this claim from an enactivist perspective.

disagreeable sensations" – practical ends.... [I]n every case the idea of expression of emotion ... enter[s] in only to confuse." (Dewey 1894, 555)

In other words, Darwin's explanations of emotional expressions don't actually appeal to emotions as such. Look again at Darwin's formulation of the first principle:

> 1. *The principle of serviceable associated Habits.* – Certain complex actions are of direct or indirect service under certain states of the mind, in order to relieve or gratify certain sensations, desires, etc.; and whenever the same state of mind is induced, however feebly, there is a tendency through the force of habit and association for the same movements to be performed, though they may not then be of the least use. (Darwin 1872, 28)

The word "emotion" appears nowhere in the principle. It is natural to assume that by "certain states of the mind," Darwin means emotions. But do those states of mind have to be emotions? They could just as well be perceptual states. Consider: A dog sees a dark, looming object. It's an enemy! The dog growls to fend it off. Over time, the dog develops a habit of growling whenever it sees a dark, looming object, even if it's not really an enemy. In those cases, the growling serves no purpose. We can apply Darwin's first principle without mentioning emotion at all. There is no need to posit a mysterious emotion to serve as a conditioned stimulus.

The second principle does not mention emotions either:

> 2. *The principle of Antithesis.* – Certain states of the mind lead to certain habitual actions, which are of service, as under our first principle. Now when a directly opposite state of mind is induced, there is a strong and involuntary tendency to the performance of movements of a directly opposite nature, although these are of no use; and such movements are in some cases highly expressive. (Darwin 1872, 28)

If seeing an enemy makes a dog tense its muscles, then seeing a friend would make a dog loosen its muscles. If, upon making up our minds, we habitually grasp, then, upon being unable to make up our minds, we may automatically toss our hands up. It is tempting to insert emotions into these examples, but we do not have to, and it is not clear that doing so adds any substance to the explanation.

The third principle follows suit:

> 3. *The principle of actions due to the constitution of the Nervous System ...* – When the sensorium is strongly excited, nervous force is generated in excess, and is transmitted in certain definite directions, depending on the connection of the nerve-cells, and partly on habit: or the supply of

> nerve-force may, as it appears, be interrupted. Effects are thus produced which we recognise as expressive. (Darwin 1872, 28–29)

Darwin claims that many animals produce loud cries when suffering intensely. But in fact, suffering explains nothing. The stimulation of the sensory apparatus produces excess nerve force, which results in loud cries. Darwin's explanation sidesteps suffering, emotion, entirely.

In sum, contemporary Darwinian theories can explain the evolutionary value of emotion. But Darwin's own theory does not and seemingly cannot. His evolutionary account of emotional expression leaves no room for an evolutionary account of emotional feeling. In bringing the former into the light, he shrouds the latter in darkness.[38]

1.5.3 Communication

The third puzzling mistake of *Expression* is that it overlooks the importance of communication. According to Darwin's first principle ("serviceable associated habits"), many expressions began as voluntary, purposive movements. If an animal comes face-to-face with a scary predator, then bristling may save its life. Bristling makes the animal appear larger and more formidable than it truly is. The predator may be deterred from attacking. The animal may survive to bristle another day. What puzzles scholars is that in documenting the original purposes of expressions, Darwin appears to overlook the role of communication entirely. Many expressions in fact originated as social signals, yet Darwin failed to see this. The first puzzle is: Why did Darwin focus on habit? The third puzzle is: Why didn't Darwin focus on communication?

In his contemporaneous review of Darwin's *Expression*, Alfred Russel Wallace (1873) applauds Darwin for pinpointing the origins of many expressions, but laments that in some cases Darwin has overlooked the most obvious explanation. For instance:

> The open mouth, and raised arms with open hands turned outwards, is an expression of astonishment very general all over the world. Mr. Darwin explains the open mouth by a complication of causes, but he omits to notice, what seems to me a very probable one, that it represents an incipient cry of alarm or fear, or call for help. (1873, 116)

[38] Drawing on Dewey (1894), Campbell (1997a) argues for this conclusion, suggesting that naturalistic explanations of emotion come up short. She elsewhere (1997b) offers an alternative theory of emotion, which somewhat resembles psychological constructionism, a theory I discuss in §4.6. I argue that Darwin's theory is closer to Campbell's alternative than it first seems.

Notice how easily Darwin's first principle could account for this expression. Suppose that whenever a person is startled, he voluntarily calls out for help. Over time, a habit may form, and this person may involuntarily call out upon being startled, even if no one is within earshot. Calling out, or even preparing to call out by opening one's mouth, would come to express astonishment. Wallace states that this sort of explanation is "a very probable one," and seems genuinely surprised that Darwin did not propose it himself.

Fifty years later, the ethologist Wallace Craig (1921) published a retrospective on Darwin's *Expression*, observing that Darwin's reluctance to attribute communicative functions to emotional expressions had aged poorly. The prevailing view in Craig's day was that expressions must have evolved to serve as social signals (Craig 1921, 361). Because Darwin denies the importance of communication, Craig concludes that *Expression* expounds a view of the evolution of emotion expression that is "utterly erroneous" (1921, 361). The social psychologist Floyd Henry Allport put the same point somewhat more diplomatically:

> [Darwin] demonstrated that facial expression, or communication, was not the original function of the facial muscles . . . It seems however that Darwin neglected the possibility of the facial movements becoming important in adaptation to the social environment. (1924, 211)

At the beginning of the twentieth century, scientists began to emphasize, more and more, the communicative functions of emotional expressions. Oskar Heinroth (1911) and Julian Huxley (1914) observed that many emotional expressions are truncations and elaborations of adaptive behaviors. A dog that is angry about sharing its food would benefit from biting and disabling its competition. However, dogs characteristically express their anger not by biting but by baring their teeth. Baring one's teeth cannot disable a competitor, but it can serve as a credible threat of biting, and it may cause the competitor to seek food elsewhere. Thus, the threat, like the act of biting, can benefit the dog by helping it to secure its meal. Huxley coins the term "ritualization" to describe the process by which serviceable actions truncate and elaborate into communicative signals, and he argues that many of the expressions that we see in animals are ritualized signals.[39]

It is especially surprising that Darwin overlooked the communicative functions of emotional expressions, given that the book he published

[39] This idea would be developed further by Nikolaas Tinbergen (1963), Amotz Zahavi (1975), and Konrad Lorenz (2002). See Burkhardt (2005) for a thorough examination.

before *Expression*, *Descent*, argues that some traits and behaviors evolved as mating signals. The peacock's large plumage does not help it to survive – indeed, it makes survival more difficult – but the large plumage does help the peacock to reproduce, since peahens are attracted to it. Thus, even though peacocks with smaller plumages are better able to escape predation and survive, they are unable to find mates, and thus are unable to pass their small plumage down to the next generation of peacocks (Darwin 1871, II:97, II:141). The large plumage is the more advantageous trait. And its advantage lies entirely in its value as a social signal. Darwin understood that traits could evolve as social signals. So why didn't he see that emotional expressions evolved as social signals? That is the third puzzling mistake of *Expression*.

In sum, scholars contend that Darwin made three mistakes in *Expression*:

1. Darwin was wrong to focus on habit.
2. Darwin was wrong to sidestep emotion.
3. Darwin was wrong to downplay communication.

These mistakes were unforced errors, given that his own theory of evolution by natural selection provided him with the knowledge he needed to correct these mistakes.

1.6 The Key to Darwin's *Expression*

My aim in this book is to solve the puzzles of *Expression* – to explain why Darwin made these so-called mistakes. What scholars have overlooked about Darwin's theory, I claim, is that it relies on an associationist theory of mind and emotion. This framing helps us to see that Darwin had good reason to focus on habit, to sidestep emotion, and to ignore communication. Darwin took this approach because he accepted the prevailing scientific theory of the mind. Darwin's "errors" were not unforced but rather forced – forced by a compelling but ultimately incorrect view of how the mind works.

Darwin's *Expression* develops an associationist theory of emotion and emotional expression. So what is associationism? Associationism is a theory in the philosophy of mind, closely related to empiricism. Empiricism is the view that all ideas derive from sense experience. (The epistemological version is that all *knowledge* derives from sense experience, but we are concerned here with the origins of all thoughts, not just those that qualify as knowledge.) Some empiricists hold that the mind begins a *tabula rasa*, or blank slate, and that sensory experiences supply the mind with content. All ideas are learned. None are innate.

If empiricism is a theory about where the mind gets its ideas, then associationism is a theory about what the mind does with these ideas once it gets them. Associationists hold that the mind ultimately does one thing: It forms associations. If you hear "salt," then you may think "pepper," since these two words are associated in your mind. Associationists analyze thinking and reasoning as "trains of thought," where each thought is connected to the next by an association. In the classic film *National Treasure: Book of Secrets*, Benjamin Franklin Gates (played by Nicolas Cage) finds the eponymous treasure by following a series of clues. One clue, found engraved on the French Statue of Liberty, reads, "These twins stand resolute." Gates ponders its meaning:

> These twins stand resolute. Let's see ... resolute twins. Resolute, and then twins. Siamese twins, Siam, trade routes between France and Thailand ... that's ridiculous. HMS Resolute, a British ship that got lost in the Arctic in the 1800s, it was salvaged by American whalers then Congress sent it back to England where the ship finally retired. Queen Victoria had two desks made from its timbers. *Voila* ... resolute twins. (Turteltaub 2007)

The word "resolute" reminds Gates of a ship, the HMS *Resolute*, which reminds him that the wood of the ship was made into two desks, called the "resolute twins," which reminds him that one of these desks sits in the Oval Office of the White House, where the next clue is located. Each step in Gates's thought process is linked by an association. Associationists seek to explain how the process unfolds by positing basic laws of association: contiguity, similarity, contrast, and so on.

Associationism is not just a theory of cognition; it is also a theory of emotion.[40] Associationists analyze emotions as trains of associations, which normally include thoughts, feelings, and actions. You see a loose dog on the street, which reminds you of the time a dog bit you as a kid, which reminds you of how much it hurt, which makes you want to get away from this dog before it can bite you. Your racing mind is in a state of fear. According to the associationist John Locke, fear is "an uneasiness of the mind, upon the thought of future evil likely to befall us" (1894, II.20.10, p. 305). The feeling and thought are linked, not intrinsically, but extrinsically, via the principle of association. We learn through repeated experience that suffering evil is painful, so the thought of evil occasions the feeling of

[40] I use the terms "emotions" and "passions" interchangeably. In §2.1 I reply to Dixon's (2003) argument that these terms belong to distinct scientific paradigms.

unease. Fear is the mind's simulation of future injury, patterned on past injuries. To be clear, Locke does not equate fear with the feeling of unease. Emotions are not mere feelings. Rather, Locke equates fear with the runaway train of ideas that barrels through the mind.[41] Fear is the feeling of unease that follows immediately and involuntarily upon the thought, or rather thoughts, of future evil. Emotions are dynamic, unfolding episodes.

To demonstrate the power of association, Locke presents a regrettably sexist example of irrational fear, composed of dubious associations, derived not from actual experience but from superstition:

> The ideas of goblins and sprights have really no more to do with darkness than light; yet let but a foolish maid inculcate these often on the mind of a child, and raise them there together, possibly he shall never be able to separate them again so long as he lives: but darkness shall ever afterwards bring with it those frightful ideas, and they shall be so joined, that he can no more bear the one than the other. (1894, II.33.10, p. 531)

The thought of darkness spawns the thought of goblins, which spawns the feeling of unease. This fear, like all fears, like all emotions, and indeed like all complex ideas, is constructed from connections among sense experiences.

Many associationists were also materialists. They viewed mental events as physical events. Thomas Hobbes identified ideas and passions with "motions." David Hartley identified them with "vibrations." Owing to their materialist perspectives, many associationists viewed emotional expressions as parts of emotions. Thus, emotions are not merely trains of thought and feeling; emotions are trains of thought, feeling, *and behavior*. The laws of association explain why the sight of a dog gives rise to the thought of danger, an uneasiness of the mind, *and the act of fleeing*. In the seventeenth and eighteenth centuries, the physiology of emotion was largely speculative. However, a crucial discovery in the early nineteenth century seemed to vindicate the associationist perspective. Sir Charles Bell (1811), Darwin's creationist foil mentioned previously, and François Magendie (1822), a French physiologist, share credit for the discovery that spinal nerves serve both motor and sensory functions.[42] More specifically, anterior spinal nerve roots send motor signals from the brain to the body, resulting in action, while posterior spinal nerve roots send sensory signals from the body to the brain, resulting in sensation. This discovery

[41] I defend this interpretation in §2.3.

[42] Cranefield (1974) argues that Bell articulated a different, incorrect view, and then falsified documents to support his claim that he had come up with the correct theory before Magendie published it.

convinced many associationists that sensation and action, emotional feeling and emotional expression, are two sides of the same nerve – er, coin.[43] The laws of association that apply to one side apply also to the other.

Think again about Darwin's first two principles: "serviceable associated habits" and "antithesis." These are associationist principles! The first states that if a sensation and action co-occur often enough, an association will form, and the occurrence of one will automatically prompt the other. (The associationist David Hartley was the first person to use the adjective "automatic" to describe involuntary human behavior.) An animal that voluntarily bristles whenever afraid will soon automatically bristle when afraid. And don't forget the facial feedback hypothesis. An animal that has built a habit of bristling when afraid may experience more intense fear when voluntarily bristling. Just as the sensation can prompt the behavior, so too can the behavior prompt the sensation, since the law of association that binds them – contiguity – is symmetrical. Darwin's second principle is based on the law of contrast. Since fear and bristling are associated, the opposite of one will prompt the opposite of the other. Even the third principle, "the direct action of the nervous system," refers to habitual associations among feelings and actions.

Darwin's aim in *Expression* is to weigh in on a hot topic in the philosophy of emotion: Can the principles of association that explain emotional feelings also explain emotional expressions? Darwin argues in the introduction that although two other associationists, Herbert Spencer and Alexander Bain, paved the way for an associationist theory of emotional expression, neither had developed their ideas sufficiently. Darwin offers *Expression* as a complete associationist theory of emotional expression, which supplements existing associationist theories of emotion.

It'll take the rest of this book to spell out the details of and defend this proposal, but I will conclude the chapter by explaining how an associationist reading of Darwin's *Expression* solves the three puzzles discussed in the preceding section.

Why does Darwin focus on habit? Because habit is the basis of all associationist theories. Repeatedly thinking of two ideas, one after the other, creates a habit, such that the thought of one will spark the thought of the other. Because associationism reduces complex mental activities to repeated iterations of a simple associative process, it fits nicely with the

[43] This idea appears most clearly in the work of Alexander Bain. I discuss Bain's theory of emotion in §3.8.2.

view that humans, with all their intelligent thoughts, are descended from decidedly less intelligent animals.

Why does Darwin sidestep emotion? Actually, he doesn't. Associationists view emotions as trains of thoughts, feelings, and actions. Emotions, on this view, are emergent events. Fear doesn't exist until the thought of danger, mental distress, and the impulse to flee become bound together in habit. Darwin asks:

> what first caused this bodily action, if the emotion was not first felt? without abrupt flush, acceleration of pulse, or rigidity of muscles, – man cannot be said to be *angry*. He may have pain or pleasure these are sensations. (2009, N 66)

Although the phrase "emotional expression" suggests that an inwardly felt emotion precedes and causes an outward expression, associationists reject this view. An inwardly felt *sensation* precedes and causes an outward expression; however, the *emotion* is the whole unfolding event – the runaway train of thought, feeling, and expression. What Darwin's readers have missed is that Darwin *does* give an evolutionary account of emotion. By explaining how a train of thought and feeling leads to action, Darwin thereby explains how emotions are constructed in evolutionary time.

Finally, why does Darwin downplay communication? Because associationists view emotions as outputs of expression as opposed to inputs. It's not that there is an emotion and then an expression, which can communicate the emotion to curious onlookers. Rather, it's that the process of producing an expression is part of the emotion. So to explain an expression in terms of a preexisting emotion is to put the cart before the giraffe.

In sum, Darwin made these "mistakes" – and only the first is really a mistake; the second and third are live ideas in the contemporary philosophy of emotion – because his associationist framework committed him to them.

I do not mean to suggest that Darwin was a mere follower. To the contrary, Darwin was excited to work within the associationist tradition because he believed that his own theory of evolution could help associationists solve a problem. Herbert Spencer put the problem like this:

> The doctrine maintained by some philosophers, that all the desires, all the sentiments, are generated by the experiences of the individual is so glaringly at variance with hosts of facts, that I cannot but wonder how any one should ever have entertained it. Not to dwell on the multiform passions displayed by the infant, before yet there has been such an amount of experience as could by any possibility suffice for the elaboration of all

> passions; I will simply point to the most powerful of all passions – the amatory passion – as one which, when it first occurs, is absolutely antecedent to all relative experience whatever. (Spencer 1855, §203)

The idea that individuals can inherit progenitors' associations solves the problem. Thus, Spencer suggests that each *species'* mind begins as a blank slate, but that members of this species could inherit the associations formed by their ancestors (Young 1970, 180–181).

Darwin saw the fruitful synthesis of associationism and evolution already in 1838, more than a decade before Spencer published his synthesis. Darwin wrote in his notebook, "Plato ... says in Phaedo that our 'necessary ideas' arise from the preexistence of the soul, are not derivable from experience – read monkeys for preexistence" (2009, M 128). Associationism says that emotions are learned. Evolutionary associationism says that emotions may have been learned by one's ancestors. Associationism explains why we express our emotions *in this way and not that.* Evolutionary associationism explains why *we* express our emotions in this way and not that.

CHAPTER 2

Darwin's Predecessors

2.1 A Forgotten Tradition in the Philosophy of Emotion

I argue that Darwin's *Expression* is a contribution to the philosophy of emotion insofar as it belongs to the associationist tradition of philosophical psychology. Much has been written about associationist theories of *cognition* in the seventeenth, eighteenth, and nineteenth centuries (Warren 1921; Young 1970; Rapaport 1974; Dacey 2020). However, virtually nothing has been written about associationist theories of *emotion* in this same period, despite their ubiquity. Associationism about emotion is, as far as I can tell, a forgotten tradition in the philosophy of emotion. My aim in this chapter is to describe this rich tradition for the first time.

Associationism developed alongside empiricism. The British empiricists – including Hobbes, Locke, Hume, Hartley, and both Mills – posited that all ideas originate in sense experience. Before you can think it, you must have seen it (or heard it or smelled it). Locke famously proposed that the mind begins as "white paper" and that only experience can add text (1894, 2.1.2, p. 121). It's easy to see how some ideas derive from experience. If I call to mind an image of my dogs, then this image may be a copy of a past perception. I saw my dogs this morning, and now I have recalled this perception, albeit more hazily, in memory. But it's harder to see how other ideas derive from experience. How do we get from sensation to the ideas of the Beautiful and the Good?

In his classic history of associationism, Howard Warren describes the short path from empiricism to associationism: "Proceeding along this line of inquiry, some principle depending on the character of experience itself is needed at the outset to account for the succession and complex union of experiences. Association appeared to furnish the requisite instrument" (Warren 1921, 10). In other words, the empiricists needed a principle – and more specifically, a principle based in experience – to account for the generation of complex ideas from simple ideas. As it happens, they gravitated to the principle of association.

The empiricists used the principle of association to explain two things: the *succession* and the *union* of ideas (Dacey 2020, §1.h). First, they used this principle to account for the sequence of ideas ("trains of thought") in consciousness. Consider Hobbes's famous account of a seeming non sequitur:

> For in a discourse of our present civil war, what could seem more impertinent than to ask (as one did) what was the value of a Roman penny? Yet the coherence to me was manifest enough. For the thought of the war introduced the thought of the delivering up the king to his enemies; the thought of that brought in the thought of the delivering up of Christ; and that again the thought of the 30 pence, which was the price of treason: and thence easily followed that malicious question; and all this in a moment of time; for thought is quick. (Hobbes 2020, III.3, p. 24; quoted in Dacey 2020, §1)

At first blush, there is no reason to ask about the value of a Roman penny during a discussion of current events. Yet Hobbes identifies a set of associations that lead naturally from the thought of civil war to the thought of a Roman penny.

Second, the empiricists used the principle of association to account for the combination of simple ideas into complex ideas. I have seen many dogs in my life. These dogs resemble each other. Granted, a three-legged Chihuahua doesn't look much like a prize-winning Greyhound, but the Chihuahua resembles a Shih-Tzu, which resembles a Terrier, which resembles a Dachshund, which resembles an Italian Greyhound, which resemble the seasoned Whippet. The principle of association – in this case, resemblance – combines my many perceptions of dogs into an abstract concept of dog.

All philosophical psychologies acknowledge the importance of association. What sets associationism apart is its insistence that *most* (if not *all*) mental processes are fundamentally associative processes.[1] The mind really only does one thing: It forms associations among sense experiences and ideas thereof. As Warren puts it, associationism seeks "to explain all complex forms of consciousness, including knowledge itself, by means of this single instrument" (1921, 10).

An alternative to empiricism is rationalism, or the view that some ideas originate in reason, independently of the senses. An alternative to

[1] Dacey (2022) notes that while associationism's most vocal critics charged it with attempting to reduce all mental activity to association, few associationists actually went that far. Hartley attempted to reduce everything to association, as did James Mill, but most other associationists limited the role of association.

associationism is faculty psychology, or the view that we must posit a set of innate mental faculties, such as emotion, reason, and will, in order to explain the activities of the mind. Some associationists deny that the faculties identified by faculty psychologists are innate. Hobbes argues that they must be learned:

> There is no other act of man's mind that I can remember, naturally planted in him, so as to need no other thing to the exercise of it but to be born a man, and live with the use of his five senses. Those other faculties, of which I shall speak by and by, and which seem proper to man only, are acquired and increased by study and industry. (Hobbes 2020, III.11, pp. 26–27)

Other associationists accept that there are innate faculties but claim that we must infer them from observable mental activity. Demeter argues that Hume adopts the latter strategy: "Instead of arguing *from* faculties, Hume argues *to* them; they are not the beginning but the aim of proper, experimental inquiry that reveals the characteristic activity of faculties" (2021, 5364).[2] Faculty psychologists seek to explain mental activities in term of mental faculties, while associationists seek to explain mental faculties in terms of mental activities.

Warren identifies four stages in the historical development of associationism (see Table 2.1). The first, "precursory period" includes Thomas Hobbes, John Locke, Bishop Berkeley, and David Hume (1921, 15). These philosophers offered the first associationist explanations of specific mental phenomena, but their theories were not general enough to explain all mental phenomena. The second stage includes David Hartley and Thomas Brown, whose theories were "fairly complete in covering the field of mental phenomena, though distorted by the crude physiology and burdened by the primitive anthropology of their time" (1921, 15). James Mill, John Stuart Mill, and Alexander Bain incorporated contemporary developments in neurology and physiology in the third stage. Finally, Herbert Spencer and George Henry Lewes sought to synthesize associationism with Darwinism by "transferring the association principle from individual to racial experience" (1921, 16). Ideas originate in experience, but perhaps not always in one's own experience. Perhaps my ideas originated in my ancestors' experiences. Mike Dacey (2020, §§3–4) observes that after its "empiricist heyday" and "fracturing" in the work of Spencer, associationism evolved into behaviorism and later connectionism. I am

[2] Demeter (2021, 5363) argues against Fodor's (1983, 123) claim that Hume denies that the mind has an innate structure. Thanks to the members of the Association Association for pushing me to clarify the differences between faculty psychology and associationism.

Table 2.1 *The historical development of associationism (Warren 1921)*

Dates	Stage	Figures
1651–1748	(1) **Precursory Period**: Early attempts to explain particular mental phenomena using the association of ideas.	Hobbes Locke Berkeley Hume
1749–1820	(2) **Mature Period**: Robust attempts to explain the mind in general using the association of ideas.	Hartley Brown
1829–1859	(3) **Integrative Period**: Attempts to update associationism in light of advances in physiology and neurology.	J. Mill J. S. Mill Bain
1855–1879	(4) **Evolutionary Period**: Attempts to synthesize associationism with evolutionary theory.	Spencer Lewes

interested here only in the associationist tradition that spans from the seventeenth through the nineteenth centuries.

Notice that Warren does not include Darwin in his list of associationists. He includes Spencer and Lewes, who incorporate Darwin's ideas into their theories. But Darwin himself is absent. One aim of this book is to demonstrate that Darwin does belong to the tradition of associationist philosophy.

In the preceding pages I have sketched a brief history of associationism. I have not yet said anything about emotion. Although histories of associationism focus on explanations of *cognition* (owing, perhaps, to the empiricists' overriding interests in epistemology), many of the associationists offered associationist explanations of *emotion* as well. In the pages to follow I will present, in broad outline, the associationist approach to emotion, followed by more detailed interpretations of individual philosophers' contributions.

But first, I should clarify and defend my use of the word "emotion." Seventeenth- and eighteenth-century philosophers favored the word "passion." Fear, anger, and sadness were *passions of the soul.* Starting in the nineteenth century, philosophers began using the word "emotion" in place of "passion," and "mind" in place of "soul." Thus, fear, anger, and sadness became *emotions of the mind.* It is tempting to treat "passions" and "emotions" as synonyms, yet Thomas Dixon (2003; 2012) argues that this is a mistake. Dixon contends that each term encompasses different mental states and carries different connotations. He suggests that the difference between passion and emotion is akin to the difference between phlogiston and oxygen (O'Grady 2005, 156–159). They are not two names for the same thing. They are theoretical constructs separated by a paradigm shift in thinking about the mind and its place in nature. Emotions came in vogue as passions fell out of fashion.

Despite Dixon's warnings, I will use the terms "passions" and "emotions" interchangeably. As I hope to show in this chapter, there is no paradigm shift between older associationist theories of the passions and newer associationist theories of the emotions. Rather, there are only gradual changes from one theory to the next. My argument is strongest in comparing the theories of David Hartley, Erasmus Darwin, and Charles Darwin. Hartley offers a theory of "the passions." Erasmus offers a remarkably similar theory of "the emotions and passions." Charles, in turn, offers a remarkably similar theory of "the emotions." The names change. But the basic theory stays the same.

What, then, are the central tenets of associationist theories of emotion? First, associationists eschew the postulation that we have an innate capacity for emotion and instead argue that emotions, like ideas, are derived from sense experiences via the principles of association. We sense the world and form associations among those sensations. We catch onto the sorts of things that cause pleasure, the sorts of things that cause pain, and the means of pursuing and avoiding them. Eventually, emotions bubble up from this process. David Hartley puts the point bluntly: "our Passions or Affections can be no more than Aggregates of Simple Ideas united by Association" (1749, 368).

Empiricist associationists, including Hobbes and Hartley, insist that emotions derive from associations formed in *one's own experience*. To be afraid of a snake, for instance, I must have had negative experiences with snakes. Evolutionary associationists, including Spencer and Darwin, allow for emotions to derive associations formed in *one's ancestors' experience*. Although I personally have not had any negative experiences with snakes, I may fear snakes given that my ancestors had negative experiences with snakes. Thus, Spencer and Darwin hold that associations acquired in one generation may become instinctual in the next.

Second, associationists analyze emotions as aggregates or sequences of simpler thoughts, feelings, and actions, linked together by associations. Alexander Bain writes that:

> the theory of Emotion will consist in the statement of the general laws that regulate the concurrence, coalition, transfer, or aggregation of the primary feelings – the Muscular states and the Sensations; and the analysis of the several Emotions will be the account of the component feelings in each, and the mode of their composition under some of these general laws. (1865, 35–36)

Fear, for instance, might be the thought of danger, the feeling of unease, and the desire to flee, occurring either together or one after the other. In fact, most associationists viewed emotions as sequences, or as "trains" of thought, feeling, and action.[3] What makes the experience an *emotion* is that the components have become so tightly linked in the mind that one cannot occur without pulling the others in tow.

This point sets associationist theories apart from many contemporary theories of emotion. Commonsensically, emotions seem to involve feelings, thoughts, and behaviors. Anger feels hot. Contempt feels cold. Joy feels exciting. Sadness feels depressing. But emotions are more sophisticated than simple itches and aches (Pitcher 1965). They seem to involve thoughts, too. Anger includes the thought of offense. Contempt includes the thought of one's own superiority. Joy includes the thought of delight. Sadness includes the thought of loss. But to really be an emotion, these thoughts and feelings have to move us, to set our bodies in motion (Scarantino 2014). Angry people clench their fists. Contemptuous people snarl. Happy people dance. And sad people cry.

Commonsensically, emotions involve feelings, thoughts, and behaviors. In the twentieth and twenty-first centuries, philosophers of emotion argued over which of these three elements is the heart, or essence, of emotion (Scarantino & de Sousa 2018). So-called *feeling theories* identify emotions with feelings, and then attempt to explain the accompanying thoughts and behaviors in relation to those feelings. Peter Goldie (2000), for instance, analyzes emotions as "feelings-toward," a special class of feelings that resemble thoughts. So-called *evaluative* or *cognitivist theories* identify emotions with thoughts, and then attempt to explain the accompanying feelings and behaviors in relation to those thoughts. Martha Nussbaum (2001) analyzes emotions as judgments about what is of pressing importance to our welfare, which get our blood moving. Finally, so-called *motivational* or *behaviorist theories* analyze emotions as dispositions toward action, and then attempt to explain the accompanying thoughts and feelings in relation to those dispositions. Andrea Scarantino (2014) argues that emotions are "behavioral programs" that compute sensory inputs into behavioral outputs.

These three approaches – feeling, evaluative, and motivational – purport to identify *intrinsic* connections between the three components of

[3] Dacey (2020, §1.h) remarks that associationists "disagreed on whether the process, force, or principle combining simple ideas into complex ideas (simultaneous association) was the same as that producing the sequences of ideas through the mind (successive association). All of the theorists discussed here accept successive association, while simultaneous association is more controversial."

emotions. By contrast, associationist theories posit only *extrinsic* connections between these components. Anger feels hot and involves thoughts of offense, not because there is any intrinsic connection between this feeling and that thought, but simply because they happened to coincide in the past. Someone got in your way. They kept you from getting what you wanted. You struck them, but they struck back harder, injuring you. Your body temperature rose as your immune system tended to the injury. Associations formed in your mind. Now, the mere thought of offense prompts the feeling of heat. Since you had also clenched your fists in the fight, the mere thought of offense may also prompt you to clench your fists, even if there is no actual opponent for you to fight.

The associationist view is that the emotion (in this case, anger) is this aggregate or train of diverse experiences – of thoughts, feelings, and actions – bound together by mental associations. Emotions are replays of formative experiences, reruns of major life events, reenactments of defining roles. They unfold automatically and, in many cases, contrary to our wills.

Associationists claim further that the aggregating/sequencing process is iterative. Although the most elementary emotions are composed of simpler, non-emotion states, these emotions can combine into compound emotions, which can aggregate further into even more complex emotions. Hobbes, for instance, analyzes indignation in terms of anger, anger in terms of courage, courage in terms of fear, and fear in terms of an aversion alongside an opinion of hurt (2020, VI.14–18, p. 44).

Third, associationists tend to be nominalists about emotion types. Although we often speak of "anger" and "fear," it would be more accurate to speak only of particular "angers" and "fears." There is no essence of anger that all episodes of anger share. Thomas Brown puts the point like this: "In nature, there are no classes. We give the names of fear, and tender pleasure, to various feelings; but there exist only particular feelings of fear, of pleasure" (Brown 1798, 321). Locke might say that there is a nominal essence of "anger," "fear," and "joy" – a convention of grouping emotions together under these names – but not a real essence – no inherent quality that makes each episode the type of emotion that it is.

It is for this reason that associationists tend to have more sprawling taxonomies of emotions. Hartley lists ten emotions. Hobbes lists more than thirty. Darwin lists more than fifty! Although we can, and sometimes do, lump diverse emotions into just a few boxes – anger, fear, sadness, and so on – we can also divvy up these emotions among a great many boxes to emphasize their subtle differences and unique manifestations. We can treat contempt, disgust, disdain, scorn, and loathing as subtly distinct categories of emotion, as Darwin does.[4] According to the associationists, taxonomies of emotions are ultimately conventional, and thus there is no reason to shy away from our rich vocabulary for emotions.

In sum, associationists view emotions as sequences of thoughts, feelings, and actions, bound together by extrinsic associations. There is no underlying essence that makes an emotion "fear" as opposed to "anger," but we have conventions for gathering diverse experiences under common labels.

My goal in this chapter is to describe this rich tradition in the philosophy of emotion. My goal in the chapters to follow is to show that Darwin's *Expression* makes a unique and important contribution to this tradition, which should secure its place in the history of the philosophy of emotion.

2.2 Thomas Hobbes

If there is one thing that everyone should know about the philosopher Thomas Hobbes (1588–1679), it's that he describes life in the state of nature as "nasty, brutish, and short." His dark and brooding account of the social contract is, for good reason, a classic of political philosophy.

If there is another thing that everyone should know, it's that Hobbes is an *empiricist*, who claims that all ideas derive from the senses. "The original of them all is what we call SENSE," he writes. "The rest are derived from that original" (2020, I.2, p. 17). For context, many of Hobbes's predecessors argued that the mind has an inherent structure – a division of mental faculties, including the intellect, the passions, and the will (Knuuttila 2023). Hobbes replies that sensation is the only original faculty, and that all others (the intellect, the passions, and the will) derive from the experiences of the senses. (2020, III.11, pp. 26–27).

The third thing that everyone should know about Hobbes is that he is a *materialist*. He claims that the universe consists entirely of matter and motion. There are no gods, no spirits, and no immaterial minds. Minds exist, to be sure, but they are material, just like everything else. A thought is a physical motion that occurs within the body. So is a feeling.

Hobbes posits that once set in motion, the mind will remain in motion until an opposing force brings it to rest (2020, II.1, p. 19). When one thought leads to another and then another, propelled by the mind's momentum, Hobbes calls the sequence a "TRAIN of thoughts" (2020, III.1, p. 23). This phrase, "train of thoughts," becomes the calling card of associationists. Hobbes, Locke, and Hartley use this phrase. Even Charles

[4] I discuss Darwin's treatment of these emotion categories in §4.4.

Darwin uses it, defining emotions as "trains of thought long associated with action" (DAR LIB 404, 333).[5]

Hobbes theorizes that trains of thought can be either guided or unguided. When unguided, thoughts can drift and combine, without much rhyme or reason. Hobbes muses that the idea of a centaur must have arisen "from the sight of a man at one time, and of a horse at another" (Hobbes 2020, II.2, p. 20). The thought of the man drifts to the thought of the horse, and these thoughts are combined in the imagination. Hobbes also gives the example of the Roman penny, quoted earlier in the chapter.

Hobbes is sometimes classified as a proto associationist, rather than a proper associationist, because he did not go so far as to identify *association* as the glue that binds the sequence together. Indeed, the words "association" and "associate" appear nowhere in *Leviathan*. He states only that ideas follow each other in time, and that the sequence is somehow based in experience.[6]

Having discussed unguided trains of thought, we turn to guided trains. Hobbes theorizes that guided trains of thought are led "by some desire and design" (2020, III.4, p. 24). The desire specifies an end, or goal. The design specifies the means of achieving that end. Suppose that you want ice cream. Your mind recalls your past efforts to acquire ice cream and then passes through the steps you took in reverse-chronological order until settling upon the first step that you need to take right now.[7] Before you got the ice cream, you had to pay the cashier. Before you paid the cashier, you had to get to the store. But first you needed to put on your pants and shoes and grab your keys. This is the point at which you stop thinking and start acting. Pants: check. Shoes: check. Keys: Where are my keys?

Hobbes defines the *passions* as the "interior beginnings of voluntary motions" (2020, VI, p. 40). In other words, the passions are those trains of thought that hurtle us, not simply toward more thoughts, but toward physical action.[8] Anger propels us toward fighting. Fear propels us toward fleeing. Grief propels us toward sobbing.

[5] Charles Darwin wrote this definition of emotion in the margins of his personal copy of Mackintosh (1837), archived in the Cambridge University Library as DAR LIB 404. Barrett quotes this marginal note in his transcription and analysis of Darwin's unpublished collection of notes titled "Old & Useless Notes about the Moral Sense & Some Metaphysical Points." See Gruber & Barrett (1974, 412n92).

[6] Kallich (1945) argues, however, that Hobbes identifies multiple principles of association.

[7] "From desire, arises the thought of some means we have seen produce the like of that which we aim at; and from the thought of that, the thought of means to that mean; and so continually, till we come to some beginning within our own power" (Hobbes 2020, III.4, p. 24).

[8] Gert (2010, 34) writes that Hobbes's "account of the passions is completely independent of his materialism." Bobier (2020, 591) argues that although Hobbes's focuses on the introspective experience of the passions, his account is consistent with his materialism. Schmitter (2021a)

Hobbes then lists more than thirty passions with pithy definitions. Here are the first five:

> For *appetite* with an opinion of attaining is called HOPE.
> The same, without such opinion, DESPAIR.
> *Aversion*, with opinion of *hurt* from the object, FEAR.
> The same, with hope of avoiding that hurt by resistance, COURAGE.
> Sudden *courage*, ANGER. (Hobbes 2020, VI.14–18, p. 44)

Scholars continue to debate Hobbes's definitions. Some say that Hobbes's definitions prove that he is a cognitivist.[9] *Cognitivists* say that emotions are composed of cognitions, or thoughts. Fear includes the thought of danger. Grief includes the thought of loss. Since Hobbes's definition of hope includes "an opinion of attaining," he seems to be saying that you can't hope for something unless you think you will attain it. On this reading, the thought is a necessary component of the emotion. These scholars read Hobbes's definitions as *conceptual analyses* – as attempts to break each emotion down into its constituent parts.

Other scholars say that Hobbes is a non-cognitivist (Bobier 2020, 592–593). *Non-cognitivists* say that emotions are not composed of thoughts. These scholars look at the sentence, "appetite with an opinion of attaining is called HOPE," and say that hope is identified with the appetite. The thought *accompanies* the emotion, but it does not *constitute* the emotion. (Compare: "the man with the funny hat is called MARK." Mark is the man, with or without the funny hat.) Perhaps the thought causes the emotion. Or perhaps the emotion causes the thought. In any case, one scholar argues that Hobbes's definitions should be read not as conceptual analyses but as *introspective guides* (Bobier 2020, 592–593). Feelings are fuzzy, and we sometimes struggle to name our emotions. Am I angry now or just hungry? On this interpretation, Hobbes suggests that we can better understand our emotions by attending to the thoughts that accompany them. We know we are experiencing hope, for instance, when we notice that our fuzzy desire for something is accompanied by the thought that we will attain it.

The problem with calling Hobbes a cognitivist or a non-cognitivist is that Hobbes does not think in these terms. It was not until the end of the eighteenth century that cognitions (thoughts) and affects (feelings) were

briefly discusses Hobbes's materialism and points out that Hobbes's view of the passions is future-oriented.

[9] Tuck (1996, 184), James (1997, 133–134), and Abizadeh (2017, 12).

clearly distinguished.[10] And it was not until the middle of the twentieth century that the cognitivism/non-cognitivism debate began in earnest (Deigh 1994). The question that Hobbes has in mind is rather this: Are the passions produced by a special, innate faculty of the mind, or do they derive, somehow, from the experiences of the senses? We could call the former position "*nativism*" and the latter position "*empiricism.*" Hobbes, the constant empiricist, argues that the passions derive from the senses.

If Hobbes derives the passions from the senses, then his definitions are neither conceptual analyses nor introspective guides. They are *blueprints*, or instructions for constructing passions out of sensations. Recall that guided trains of thought are led "by some desire and design" (Hobbes 2020, III.4, p. 24).[11] And notice that each passion is defined in terms of a desire and a design. Hope is appetite plus opinion of attaining. Fear is aversion plus opinion of hurt. The passion is what you get when you fasten the desire to the design.

Let's investigate hope in more detail. Experience teaches us to desire good things. Experience also teaches us the means of attaining those things. One day, you see a hiring notice for your dream job. You read the description and realize that you are the perfect candidate. Your imagination runs wild. The hiring committee is rightfully impressed by your application. They fly you out for an interview. You knock it out of the park. They call you with an offer. You immediately accept. Everyone lives happily ever after. This train of thought, guided by a desire and a design, is what we call hope. Hobbes's goal is to show that no special faculty of emotion is needed. Hope is the great-grandchild of sensation. From sensations come desires. From desires come designs. And from desires and designs come the passions. (Hobbes also attempts to derive the will from the passions, making it the great-great-grandchild of sensation.[12])

[10] See my discussion of Hume below (§2.4).

[11] In the earlier *Elements of Law*, Hobbes defines the passions more simply in terms of designs. Hope, for instance, is an "expectation of good to come" (1640, 9.52). Bobier (2020, 594) offers an illuminating analysis of Hobbes's view in the *Elements*.

[12] Hobbes describes "deliberation" as a sequence of opposing passions, and "the will" as the final passion in that sequence, which leads directly to action (2020, VI.53, p. 47). Imagine that a young child sees a dog on the street. The child thinks immediately of two other dogs: her beloved family dog and the neighbor's dog that bit her. Thinking of the family dog, she remembers how pleasant it is to scratch a dog behind the ears. Thinking of the neighbor's dog, she remembers how painful it is to be bitten. The child wants the pleasure of scratching this dog on the street. But she also wants to avoid the pain of being bitten. She experiences hope, but also fear. Her mind jumps from one to the other and then back to the first. Ultimately, her hope proves stronger, and as her fear decays, she walks toward the dog to scratch it behind the ears.

There is a puzzle game for toddlers called Chugga Choo. It includes puzzle pieces shaped like the parts of a train. The locomotive, or front of the train, has only one connector at its back. The caboose, or back of the train, has only one connector at its front. The cars, or middles of the train, have connectors on their fronts and backs. You can build trains of any length, ranging from a short two-section train (locomotive, caboose) to a long fourteen-section train (locomotive, twelve cars, caboose).

As an illustration of Hobbes's theory, the locomotive represents a desire. The caboose represents an action. The cars represent the train of thought that leads from the desire to the action. A simple train might lead from a desire, through a design, to an action. A more complicated train might lead from a desire, through multiple designs and other desires, to an action. Look again at the first five passions on Hobbes's list:

> For appetite with an opinion of attaining is called HOPE.
> The same, without such opinion, DESPAIR.
> Aversion, with opinion of hurt from the object, FEAR.
> The same, with hope of avoiding that hurt by resistance, COURAGE.
> Sudden courage, ANGER. (Hobbes 2020, VI.14–18, p. 44)

Courage is an aversion to danger, followed by an opinion of hurt, followed by an opinion of attaining safety through resistance. The train terminates in the act of resistance. When the train picks up speed, barreling down the tracks, the result is not courage, but explosive anger.

As I mentioned earlier, Hobbes may not have identified associations as the links connecting each thought in a train of thought (cf. Kallich 1945). But he set the foundation for future associationists by analyzing passions as trains of thought leading to action (see Table 2.2). Darwin is doubly indebted to Hobbes. Darwin's account of the struggle for existence that drives natural selection is inspired (indirectly) by Hobbes's account of the "nasty, brutish, and short" existence that drives people to form a social contract. And Darwin's definition of emotions as "trains of thought long associated with action" (DAR LIB 404, 333) is inspired (also indirectly) by Hobbes's pioneering theory of the passions.

2.3 John Locke

The philosopher John Locke (1632–1704) introduces the language of "association" into the empiricist tradition. In the fourth edition of *An Essay Concerning Human Understanding* (1700), he adds a chapter called

Table 2.2 *Hobbes's emotions*

EMOTION	Element 1	Element 2	Element 3	Element 4	*Modifier*
HOPE	Appetite	Opinion of attaining			
DESPAIR	Appetite	Opinion of not attaining			
FEAR	Aversion	Opinion of hurt			
COURAGE	Aversion	Opinion of hurt	Opinion of avoiding hurt by resistance		
ANGER	Aversion	Opinion of hurt	Opinion of avoiding hurt by resistance		*Sudden*
CONFIDENCE	Appetite	Opinion of attaining			*Constant*
DIFFIDENCE	Appetite	Opinion of not attaining			*Constant*
INDIGNATION	Aversion	Opinion of great hurt to another	Opinion of avoiding hurt by resistance		
LOVE	Love (of one singularly)	Desire (to be singularly beloved)			
JEALOUSY	Love (of one singularly)	Desire (to be singularly beloved)	Aversion	Opinion that love is not mutual	
GLORY	Joy	Imagination of one's own power and ability			
DEJECTION	Grief	Opinion of want of power			
LAUGHTER	Joy	Imagination of one's own power and ability			*Sudden*
WEEPING	Grief	Opinion of want of power			*Sudden*
PITY	Grief	Opinion of calamity of another			
COMPASSION	Grief	Opinion of calamity of another	Imagination that calamity may befall oneself		
ENVY	Grief	Opinion of the success of a competitor	Desire (to hinder the competitor)		

"Of the Association of Ideas." It begins with a distinction between two types of connections between ideas:

> Some of our ideas have a *natural* correspondence and connexion one with another: it is the office and excellency of our reason to trace these, and hold them together in that union and correspondence which is founded in their peculiar beings. Besides this, there is another connexion of ideas wholly owing to *chance* or *custom*: ideas, that in themselves are not all of kin, come to be so united in some men's minds, that it is very hard to separate them; they always keep in company, and the one no sooner at any time comes into the understanding, but its associate appears with it; and if they are more than two, which are thus united, the whole gang, always inseparable, show themselves together. (1894, II.33.5, p. 529)

A train of thoughts connected by "natural correspondence[s]" is rational and right. For instance, smoke and fire have a natural correspondence, so the inference from smoke to fire is justified. A train of thoughts connected by "chance or custom" is irrational and wrong. For instance, broken mirrors and bad luck have no natural correspondence, so the inference from a broken mirror to seven years of bad luck is unjustified.

Later associationists would try to explain *both* types of connections using the laws of association. If you repeatedly witness smoke and fire together, then these ideas will become linked in your mind. The next time you see smoke, you will think of fire. However, Locke explains only the *second* type of connection ("chance or custom") using the laws of association. If you are repeatedly told that broken mirrors bring bad luck, then these ideas will become linked in your mind. The next time you break a mirror, you will think that bad luck will follow. For Locke, the phrase "the association of ideas" connotes prejudice and irrationality (Tabb 2019, 77, 91–97). For later associationists, it would denote the fundamental operations of the mind.

Locke's definitions of the passions, like Hobbes's, are deceptively simple (see Table 2.3). He defines joy as "a delight of the mind, from the consideration of the present or assured approaching possession of a good" (1894, II.20.7, p. 305). You see a round, shiny object on the ground. You bend down for a closer look. It's a quarter! Upon recognizing that you have found something good, you experience delight. Locke defines fear, in turn, as "uneasiness of the mind, upon the thought of future evil likely to befall us" (1894, II.20.10, p. 305).[13] You are hiking in the woods when a bear

[13] Sodano (2017) argues that Locke's analysis of some passions in terms of "unease" signals an important shift from older analyses of passions as "disease."

Table 2.3 *Locke's emotions*

EMOTION	Opinion	Affect
JOY	Consideration of the present or assured approaching possession of a good	Delight
SORROW	The thought of a good lost	Uneasiness
HOPE	The thought of a profitable future enjoyment	Pleasure
FEAR	The thought of future evil likely to befall us	Uneasiness
DESPAIR	The thought of the unattainableness of any good	Uneasiness/ indolency
ANGER	The thought of receiving injury and the intent to seek revenge	Uneasiness
ENVY	The consideration of a good we desire, obtained by one we think should have had it before us	Uneasiness

suddenly appears before you. If you think that the bear will attack you, then you will immediately feel uneasy.

Locke's definitions follow a simple pattern: A passion is a feeling following a thought. However, there are two ways of interpreting these definitions. According to a *narrow-scope* interpretation, Locke identifies the passion with the feeling only. Joy is a feeling of delight. Fear is a feeling of uneasiness. Anger is a qualitatively different feeling of uneasiness. Although these feelings are caused by thoughts – joy, by the thought of possessing something good; fear, by the thought of an impending evil; anger, by the desire to seek revenge for an injury – these thoughts are not parts of the passions. They are the causes of the passions, their antecedents.

According to a *broad-scope* interpretation, Locke identifies the passion with the feeling–thought combo. Joy is the train of thought that begins with the thought of attaining something good and ends with a feeling of delight. Fear is the train of thought that begins with the thought of an impending evil and ends with a feeling of uneasiness. Anger is the thought of receiving injury, the intent to seek revenge, and the feeling of uneasiness.

Some scholars have favored the narrow-scope interpretation. Indeed, Locke's theory is sometimes dismissed as a crude form of non-cognitivism, which equates emotions with simple, ineffable feelings (e.g., Scarantino & de Sousa 2018, §3).

However, two considerations weigh in favor of the broad-scope interpretation. First, most of Locke's definitions take the same form: A passion is a feeling following a thought. But one definition takes the reverse form: A passion is a thought followed by a feeling. Locke defines despair as "the thought of the unattainableness of any good, which works differently in men's minds, sometimes producing uneasiness or pain, sometimes rest or indolency" (1894, II.20.11, pp. 305–306). If Locke identified passions with mere feelings, then he would have defined despair as "uneasiness or indolency, depending on men's minds, from the thought of the unattainableness of any good." The fact that he doesn't suggests that the two formulas – *a passion is a feeling following a thought* and *a passion is a thought followed by a feeling* – differ only in style, not in substance. In every case, the passion is the sequence of thought and feeling.

The second reason to favor the broad-scope interpretation is that it allows us to make sense of Locke's associationist analysis of irrational passions. This analysis occurs right after Locke's distinction between two types of connections between ideas (natural versus merely customary). He argues that an emotion is rational if the link between the thought and the feeling is natural but irrational if the link between the thought and the feeling is merely imagined. I already mentioned his analysis of the child's fear of the dark, instilled by the superstitions of a "foolish maid" (1894, II.33.10, p. 531). In this case, the threat posed by the monsters under your bed and in your closet is in your head.

Locke's second example involves hatred. It is reasonable to be upset if someone accidentally hurts you. However, it is irrational to hate someone based on that accident. Locke imagines how a person can stew on a slight until the mere thought of a person is accompanied by extreme displeasure:

> A man receives a sensible injury from another, thinks on the man and that action over and over, and by ruminating on them strongly, or much, in his mind, so cements those two ideas together, that he makes them almost one; never thinks on the man, but the pain and displeasure he suffered comes into his mind with it, so that he scarce distinguishes them, but has as much an aversion for the one as the other. Thus hatreds are often begotten from slight and innocent occasions, and quarrels propagated and continued in the world. (1894, II.33.11, p. 532)

What makes the hatred irrational is that there is no natural correspondence between the thought of the person and great displeasure. The link is forged through habit.

Locke's final example is of dread. He describes the irrationality of dreading a place, simply because it is connected in one's mind with illness and death:

> A man has suffered pain or sickness in any place; he saw his friend die in such a room: though these have in nature nothing to do one with another,

> yet when the idea of the place occurs to his mind, it brings (the impression being once made) that of the pain and displeasure with it: he confounds them in his mind, and can as little bear the one as the other. (1894, II.33.12, p. 532)

As in the previous examples, what makes the passion irrational is that it consists of an unnatural pairing of thought and feeling.

In sum, Locke's view is that a *passion* occurs when a thought and a feeling have become bound together in experience, such that the thought automatically triggers the feeling. The feeling is forced upon us, as it were, making the experience passive, and thus a "passion."

Locke's other important contribution to the associationist tradition is the concept of a "nominal essence," or a conventional categorization. We should not be concerned with the "real essence" of fear or anger. Instead, we should be concerned only with our conventional ways of categorizing emotions. We tend to call a passion "fear" when it includes the thought of impending danger, but it would be a mistake to think that this thought is the real essence of danger. Conceptual analysis can give us nominal essences, which are somewhat useful to have, but conceptual analysis should not be the central project of the philosophy of emotion. Rather, the central project is to explain how the thought and feeling are conjoined.

2.4 David Hume

David Hume (1711–1776) made several important contributions to the associationist tradition.[14] However, his theory of emotion differs in important respects from Hobbes's and Locke's. Whereas they analyze emotions as trains of ideas, Hume analyzes emotions as simple feelings, which may however be associated with their causes and effects. For this reason, I classify Hume as a quasi-associationist about emotion.

Hume's first contribution is the distinction between thoughts and feelings. Locke had used the generic term "idea" to refer to all manner of thoughts and feelings, without drawing a clear distinction between them. Even pleasures and pains are "ideas." Hume begins his *Treatise* by dividing mental states into two categories: "impressions" and "ideas" – roughly, feelings and thoughts (2007, 1.1.1.1, p. 7). He distinguishes them based on their "force and liveliness" (2007, 1.1.1.1, p. 7). Impressions are more

[14] See Warren (1921, 43–46), Broadie (2003, 66–69), and Dacey (2020, §1.b) for brief discussions of Hume's place in the associationist tradition.

vivid and lively than ideas. All ideas, however, are copies of impressions. These copies just happen to be fainter. Think of those old photocopy machines you could sit on, whose muddied "copies" were hardly deserving of the name.

Impressions divide into "impressions of sense," including perceptions, and "impressions of reflection," including emotions and passions (2007, 1.1.2.1, p. 11). If you touch a hot stove, then you will experience a sensory impression: pain. If you think of touching a hot stove, then you will experience a reflective impression: fear. The idea of touching a hot stove is a faint copy of the original sensory impression you experienced upon touching the hot stove. Thus, emotions and passions are the children of ideas and the grandchildren of sensations.

Hume at times uses the term "emotion" to denote relatively *calm* reflective impressions, including those of aesthetic appreciation, and "passion" to denote relatively *violent* reflective impressions, including love and hate (2007, 2.1.1.3, p. 181).[15] But at other times he contrasts "calm passions" with "violent emotions" (2007, 2.3.4.9, p. 268).[16] The difference between emotions and passions, if there is one, is not entirely clear.[17]

Hume divides the passions into "direct" and "indirect" passions (see Table 2.4). Direct passions include joy, grief, desire, and fear. These feelings "arise immediately from good or evil, from pain or pleasure" (Hume 2007, 2.1.1.4, p. 181). Indirect passions include love, hate, pride, and humility. These feelings arise from the association of ideas.[18] (More on this shortly.) Although the indirect passions arise from the association of ideas, they are nonetheless "simple and uniform impressions" (Hume 2007, 2.1.2.1, p. 182). We cannot introspect their internal compositional structure.[19]

Hume's second contribution to the associationist tradition is his account of association. He claims that some associations are based on *resemblance*, others on *contiguity*, and still others on *cause and effect* (2007, 1.1.4.2, p. 13). Locke had introduced the term "association," but he said only vaguely that it operates via "chance or custom" (1894, II.33.5, p. 529). Hume spells out the operations of association with greater precision.

[15] Dixon (2003) emphasizes Hume's distinction between emotions and passions to support his claim that "emotion" and "passion" are not synonyms.

[16] McIntyre (2000) offers an insightful analysis of Hume's account of the calm passions.

[17] See Fieser (1992, 9–12) for some additional difficulties in interpreting Hume's taxonomy of passions.

[18] E.g., Hume (2007, 2.1.2, pp. 182–184). See Davidson (2001), Merivale (2009), and Schmitter (2021b, §5) for critical examinations.

[19] Thanks to Tamás Demeter for correcting several of my misunderstandings of Hume. See Demeter (2021, 5366).

Table 2.4 *Hume's emotions (Fieser 1992)*

Direct Passions		
Primary	Secondary	Indirect Passions
Revenge	Joy	Love
Sympathy	Grief	Hate
Hunger	Desire	Pride
Lust	Fear	Humility

First, ideas and impressions can be associated on the basis of resemblance. You see a dog on the street. It reminds you of a similar looking dog that bit you when you were a kid. The impression of sense leads to an idea via the principle of resemblance.

Second, ideas can be associated on the basis of contiguity, occurring together in time and/or space. When you remember the dog that bit you, you may also recall your pain, since the sight of the dog and the pain from its bite occurred at the same time and place. The idea (a copy of an impression of sense) leads to another idea (a copy of another impression of sense) via the principle of contiguity.

Third, ideas can be associated on the basis of cause and effect. If yelling at dogs has caused them to retreat in the past, then you may yell at the dog you see now, expecting it to retreat. In this case, the ideas are connected via the principle of cause and effect.

Using these principles, Hume offers sophisticated associationist analyses of the indirect passions. Pride, for instance, results from an association between the pleasure one derives from one's own characteristics and oneself (2007, 2.1.2, pp. 182–184; Davidson 2001). More specifically:

1. Smith has an initial idea of some possession (or subject), such as a new suit of clothes, and this idea produces pleasure.. . .
2. This feeling of pleasure is then immediately *associated* with a feeling of pride owing to the resemblance of these two distinct, but pleasurable feelings. (This association of impressions is the first of the two associations involved in the double relation.)
3. This feeling of pride then causes Smith to have an idea of himself (as the object of pride).
4. The idea of himself is then *associated* with the idea of his new clothes (the "subject" of his pride). This association of ideas (which completes the double relation) might use, for instance, the connecting principle of

> contiguity: Smith's idea of himself would naturally lead him to consider his surrounding possessions, here, his clothes. (Fieser 1992, 8)

Analogously, love results from an association between the pleasure one derives from another's characteristics and that person (2007, 2.2.1, pp. 214–216; Schmitter 2021b, §5).

In contrast to Hobbes, Locke, and later associationists, Hume does not identify passions with complex trains of ideas. Instead, he identifies them only with the impressions of reflection that occur within these trains of ideas. In other words, although pride is connected to other impressions and ideas, via the principles of association, "pride" denotes a simple and uniform pleasurable feeling, which cannot be introspectively decomposed into parts. In the numbered list above, pride is the simple feeling identified in (2) and (3). For this reason, many scholars regard Hume as the quintessential non-cognitivist, who identifies passions with lively feelings, not representational thoughts.[20]

In sum, Hume leaves us with a useful distinction between thoughts and feelings, as well as three principles of association: contiguity, resemblance, and cause and effect. Furthermore, Hume offers sophisticated associationist analyses of indirect passions, including love, hate, pride, and humility. However, Hume does not follow other associationists in identifying the passions with trains of thoughts, feelings, and actions. For Hume, passions are just feelings.

2.5 David Hartley

If you look up the word "associationist" in a dictionary, then you will find a picture of David Hartley (1705–1757).[21] Mike Dacey writes that "Hartley's most important contribution is the very project of describing an entire psychology in associative terms" (2020, §1.c). Joan Walls calls Hartley's magnum opus, *Observations on Man* (1749), "the synthesis and origin of psychological associationism" (1982, 260). Hobbes set the foundation by supposing that all ideas originate in the senses. Locke added the

[20] See, for instance, Hume (2007, 2.1.2.1, p.182). Kenny (1963, 20), Pitcher (1965), Lyons (1980, 10), and Fieser (1992) categorize Hume in these terms. However, Collier (2011) challenges this interpretation, and offers a richer interpretation of Hume's "science of emotions." Similarly, Merivale (2009) argues that Hume revised his account of indirect passions, changing them from simple impressions to complex states.

[21] Scholarly treatments of Hartley's associationism include Oberg (1976), Walls (1982), Buckingham & Finger (1997), Mischel (1966), Koschorke (1999), Thébert (2014), and Walsh (2017). Dacey (2020) provides a helpful overview.

frame by proposing that some ideas are formed by association. But it was Hartley who built the house and lived in it.

Walls refers to *Observations* as a "synthesis" because of its innovative attempt to combine Newton's theory of vibration with Locke's theory of association (Hartley 1749, 5). Newton suggested at the end of his *Principia* (1687) that his laws of motion could explain the actions of human nervous system:

> all sensation is excited, and the members of animal bodies move at the command of the will, namely, by the vibrations of this Spirit, mutually propagated along the solid filaments of the nerves, from the outward organs of sense to the brain, and from the brain into the muscles. But these are things that cannot be explained in few words, nor are we furnished with that sufficiency of experiments which is required to an accurate determination and demonstration of the laws by which this electric and elastic Spirit operates. (Newton 1803, 314; quoted in Wade 2005)

Hartley accepted Newton's invitation to study the laws by which the nervous system operates.[22] He proposes that external stimuli cause nerves to vibrate. More specifically, he proposes that these stimuli cause tiny particles in the nerves to vibrate. "External Objects impressed upon the Senses occasion, first in the Nerves on which they are impressed, and then in the Brain, Vibrations of the small, and, as one may say, infinitesimal, medullary Particles" (1749, 11). These vibrations travel to the brain. Through repeated experience, the brain can learn and reproduce the vibrations it receives.

The doctrine of vibration explains the physical actions of the nervous system. Locke's doctrine of association explains the psychological actions of the mind. Corresponding to every vibration in the nerves is a sensation or idea in consciousness. As the brain learns to reproduce past vibrations, so too does the mind recall past sensations. Hartley struggles to articulate the precise relationship between physical vibrations and psychical associations, but he resists the dualistic picture of mental states interacting causally with physical states:

> If that Species of Motion which we term Vibrations, can be shown, by probable Arguments, to attend upon all Sensations, Ideas, and Motions, and to be proportional to them, then we are at Liberty to make Vibrations the Exponent of Sensation, Ideas, and Motions, or these the Exponents of

[22] It falls outside the scope of my project to compare Hartley's theory of the passions with Descartes's, Spinoza's, and Malebranche's theories, which likewise take up Newton's invitation. See Hoffman (1991) for an insightful discussion of these latter theories.

> Vibrations ... however impossible it may be to discover in what Way Vibrations cause, or are connected with Sensations, or Ideas; i.e. tho' Vibrations be of corporeal, Sensations and Ideas of a mental Nature (Hartley 1749, 33–34; quoted in Walls 1982, 268)

Famously, Hartley posits only one principle of association: contiguity. As Walls puts it: "Contiguous associations, simultaneous or successive, are the means by which mental and physical particulars are mechanically held together and organized" (Walls 1982, 269–270). Likewise, Dacey writes that "Hartley only posited association by contiguity and did not allow for any other laws of association" (Dacey 2020). Just as mental associations are guided by the principle of contiguity, so too are nerve vibrations:

> Any Vibrations A, B, C, etc., by being associated together a sufficient Number of Times, get such a Power over a, b, c, etc., the corresponding miniature Vibrations, that any of the Vibrations A, when impressed alone, shall be able to excite b, c, etc., the Miniatures of the rest. (Hartley 1749, 67; quoted in Walls 1982, 270).

Events that occur together in time become connected in the mind, since nerves that vibrate together eventually align.

With this outline of Hartley's general project behind us, we turn to his theory of the passions. The theory begins with the observation, quoted previously, that "our Passions or Affections can be no more than Aggregates of Simple Ideas united by Association" (1749, 368).[23] Like Hobbes and Locke, Hartley views each emotion as a train of experiences, each linked to the next by an association. What are those experiences – the individual cars making up the train?

Hartley suggests that every emotion involves either pleasure or pain. As such, the most general classification of the emotions is in terms of love and hatred:

> As all the Passions arise thus from Pleasure and Pain, their first and most general Distribution may be into the Two Classes of Love and Hatred; i.e. we may term all those Affections of the pleasurable Kind, which Objects and Incidents raise in us, Love; all those of the painful Kind, Hatred. (Hartley 1749, 369)

Drinking beer causes pleasure, so the thought of drinking beer occasions the feeling of love, which is the thought of experiencing that pleasure

[23] Warren's (1921, 60) otherwise detailed and thorough analysis of Hartley's associationism says only this about Hartley's theory of the passions. In what follows, I supplement Warren's analysis with a more detailed exploration of Hartley's theory of the passions.

again. Drinking medicine causes pain, so the thought of drinking medicine occasions the feeling of hatred, which is the thought of experiencing that pain again.[24]

Next, we associate pleasures with past actions that secured those pleasures, and we associate pains with past actions that incurred those pains. For instance, we associate beer with going to the pub, and we associate medicine with going to the doctor. Hartley claims, "When our Love and Hatred are excited to a certain Degree, they put us upon a Variety of Actions, and may be termed Desire and Aversion" (1749, 369). In other words, the love of beer prompts the desire to go to the pub, and the hatred of medicine prompts the aversion to going to the doctor.

Desire and aversion, like love and hatred, are learned via the principle of association:

> The young Child learns to grasp, and go up to the Plaything that pleases him, and to withdraw his Hand from the Fire that burns him, at first from the Mechanism of his Nature, and without any deliberate Purpose of obtaining Pleasure, and avoiding Pain, or any explicit Reasoning about them. By degrees he learns, partly from the Recurrency of these mechanical Tendencies, inspired by God, as one may say, by means of the Nature which he has given us; and partly from the Introduction and Imitation of others; to pursue every thing which he loves and desires; fly from every thing which he hates; and to reason about the Method of doing this, just as he does upon other Matters. (1749, 369–370)

The passage is reminiscent of Hobbes. We are born with the ability to sense the world, to store copies of those sensations, and to form associations among those copies. By this process alone, the passions bubble up and coagulate.[25]

The train of associations does not end here. If I associate the pleasure of drinking beer with the pub, then I will associate it also with the walk to the pub, with the various landmarks I pass on the way, and even with Thursdays, the day I normally visit the pub. My desire for the pleasure of drinking beer becomes a desire for all the steps that lead me to my

[24] Notice that while Hume analyzes love and hatred as indirect passions, derived from simpler passions via association, Hartley analyzes love and hatred as the simplest passions, from which others derive via association.

[25] Hartley argues further that general desires and aversions are derived from particular desires and aversions. In keeping with his empiricist, associationist approach, he asserts that "the most general of our Desires and Aversions are factitious; i.e. generated by Associations" (1749, 370). It's not that we have a general desire for happiness, which manifests in particular desires for pleasant things. Rather, we have a bunch of particular desires for pleasant things, which conglomerate into what we conventionally call a general desire for happiness.

beloved beer. As the sequence of steps becomes longer and longer, however, the probability that I'll get all the way from the first step to the last becomes lower and lower. As I think of Thursdays, I *hope* that I'll drink a beer, but I *fear* that I won't. So many things can get in the way. Hartley writes:

> Hope and Fear are . . . the Attendants upon Desire and Aversion. These affect us more or less, according to the more or less frequent Recurrency of the pleasing and painful Ideas, according to the greater or less Probability of the expected Event, according to the greater or less Distance of Time, etc.; the Power of Association displaying itself every-where in the Agitations of the Mind excited by these Passions. (1749, 372–373)

When we get what we hoped for, we experience *joy*. When we get what we feared, *grief*. When joy passes, we experience a lingering *recollection*. When grief passes, a lingering *resentment* (Hartley 1749, 373). In every case, the passion is the mind's simulation of the pursuit of pleasure or the avoidance of pain, passing through the various steps imprinted in our minds (see Table 2.5).

Hartley concludes that these ten emotions – five "grateful passions" and five "ungrateful passions" – span the full range of human emotionality (1749, 373).

Associations are not limited to ideas. As the nerves vibrate, ideas lead to actions – to expressions. Love is a feeling of pleasure followed by an embrace. Aversion is a feeling of pain followed by withdrawal. Joy, we all know, is expressed in a smile. Grief in tears. Consistent with his materialist leanings, Hartley views passions as trains of associations that include feelings, thoughts, *and actions*. This fits with a general conception

Table 2.5 *Hartley's emotions*

EMOTION	Idea 1	Idea 2	Ideas 3+	*Temporality*
LOVE	Pleasure			
DESIRE	Pleasure	Proximate action		
HOPE	Pleasure	Proximate action	Distal action(s)	*Future*
JOY	Pleasure	Proximate action	Distal action(s)	*Present*
RECOLLECTION	Pleasure	Proximate action	Distal action(s)	*Past*
HATRED	Pain			
AVERSION	Pain	Proximate action		
FEAR	Pain	Proximate action	Distal action(s)	*Future*
GRIEF	Pain	Proximate action	Distal action(s)	*Present*
RESENTMENT	Pain	Proximate action	Distal action(s)	*Past*

of emotions as complex states with characteristic affective, cognitive, and behavioral components.

As we shall see in §§3.2 and 3.3, Darwin's theory of emotion is heavily influenced by Hartley, both directly and indirectly through Hartley's influence on Darwin's grandfather, Erasmus Darwin.

2.6 Later Associates

My aim in this chapter has been to describe the associationist tradition in the philosophy of emotion – a tradition that has largely been forgotten. I presented the development of this tradition in the works of Hobbes, Locke, Hume, and Hartley. If associationism was nascent in the first three, it reached maturity in the last.

In the chapters to follow, I will present several other associationist theories of emotion in more detail. Here I will mention them only in passing.[26]

Shortly after Hartley's death, one of his staunchest defenders, Joseph Priestley (1733–1804), published an abridged edition of Hartley's *Observations* retitled *Hartley's Theory of the Human Mind* (1775). This edition excised Hartley's theory of nerve vibrations, which even Priestley found unconvincing. By making Hartley's work more accessible and plausible, Priestley helped it to find a new generation of readers.[27]

Priestley joined the Lunar Society of Birmingham and brought his passion for Hartley with him. This passion rubbed off on one of the society's founders, Erasmus Darwin (1731–1802), Charles's grandfather. In his *Zoonomia* (1796), Erasmus defended Hartley's associationist theory of mind and emotion (Warren 1921, 68), and even developed a novel associationist analysis of emotional expression. Erasmus proposed that if a person voluntarily performs an action while experiencing a feeling, an association may form. The next time the person experiences that feeling, they may involuntarily perform that same action. At this point, the action has become an expression of the feeling. Erasmus's principle rubbed off on Charles, who included it in *Expression* as his first principle of expression.

Erasmus also suggested that opposite feelings will become associated with opposite expressions. He observes that when an infant experiences the pain of hunger, it puckers its lips. When the infant is offered a breast, and

[26] I will not discuss two minor associationists, Abraham Tucker (see Warren 1921, 65–67) and Archibald Alison (see Warren 1921, 68).

[27] See Townsend (1993) for an analysis of Hartley's influence on Priestley.

experiences the pleasure of a full stomach, its lips pull in the opposite direction. The pucker draws back into a smile. Charles adopted this insight as his second principle of expression.

A young upstart, Thomas Brown (1778–1820), wrote a fierce critique of Erasmus's *Zoonomia* that catapulted his career.[28] By the second decade of the nineteenth century, he was widely considered *the* authority on the philosophy of mind and emotion. Brown's *Lectures on the Philosophy of the Human Mind* (1824) went through twenty editions in forty years. Inspired by Locke, he argues that emotion terms range over diverse experiences without any shared essence. Brown presents a sprawling taxonomy of emotion that resembles Charles Darwin's even more sprawling taxonomy from *Expression.*

James Mill (1773–1836), the less famous father of John Stuart Mill and friend of Jeremy Bentham, defended Hartley's theory against Brown's objections in his *Analysis of the Phenomena of the Human Mind* (1829). In his summary of Mill, Warren notes that Mill succeeds in reducing the passions to

> sensations and ideas, combined into groups, or following in trains, by the single process of association; and the principle of association itself is reduced to its simplest terms – the tendency of ideas to group themselves or succeed one another after the manner of their original. (Warren 1921, 94)

Mill's theory adds little substance, but much clarity and systematicity, to Hartley's foundation.

Alexander Bain (1810–1877) published two major works on associationism.[29] *The Senses and the Intellect* (1855, 1864) offers an associationist analysis of cognition. *The Emotions and the Will* (1859, 1865) offers an associationist analysis of emotion. He also edited, alongside John Stuart Mill (1806–1873), a critical edition of James Mill's *Analysis of the Phenomena of the Human Mind.* Just as Hartley attempted to develop a theory of the physical basis of association, Bain also dedicates considerable focus to the structure and function of the brain. Whereas Hartley's analysis was outdated even before it was published, Bain's analysis was at the bleeding edge of neurophysiology. One of Bain's most important principles, "the law of diffusion," would influence Darwin's first and third principles of expression.

[28] For scholarly treatments of Brown's associationism, see Mackintosh (1837), Warren (1921), Landes (1926), Dixon (2003), and Dacey (2015).

[29] See Young (1970), Greenway (1973), Kennedy (2012), Harper (2019), and Dacey (2022) for analyses of Bain's associationism.

Finally, Herbert Spencer (1820–1903) developed an associationist analysis of mind and emotion in his *Principles of Psychology* (1855) and an associationist analysis of emotional expression in "The Physiology of Laughter" (1860). Spencer's theory resembled Bain's, with the notable exception that Spencer's theory accords an important role to evolution.[30] It was Spencer, not Charles Darwin, who coined the phrase "survival of the fittest" to describe natural selection. Spencer's theory also inspired Darwin's third principle of expression.

This whirlwind history brings us to Charles Darwin (1809–1882). The next chapter presents Darwin's engagements with these associationists and argues that Darwin conceived *Expression* as the next entry in the associationist canon.

[30] See Young (1970) and Kennedy (2012) for discussions of Bain's relation to Spencer.

CHAPTER 3

Darwin's Associationism

3.1 Darwin and the Associationist Tradition

Building on the work of Hobbes, Locke, and Hume, David Hartley (1749) constructed an associationist theory of emotion. He claimed that emotions are trains of thoughts and feelings, connected by associations, resulting in action. You see a wild dog on the street. It reminds you of the time a dog bit you. You remember how much it hurt. Wanting to avoid that pain again, you turn and run away. The whole sequence of events – the sight of the dog, the thought of the past, the aversion to pain, and the attempt to flee – constitutes fear.

Charles Darwin was keenly interested in Hartley's theory of emotion. His early notebooks on expression contain the following remarks:

> I think there is much truth in [Hartley's] doctrine, for we can thus explain love of place. – although here we have not received pleasure *from* the place, but merely *in* the place. & yet place calls up pleasure. (Darwin 2009, OUN 49–50)

> But does not Hartley['s] explanation apply perfectly to the origin of these instincts[?] (Darwin 2009, OUN 50)

> I should think some parts of the emotive part of man, may be quite artificial, as avarice love of *gold.* – love of fame – Yes Hartley explains this. (Darwin 2009, OUN 54)

Darwin was interested in Hartley's theory because it fills a gap in his own theory of evolution. How did the human mind, with all its profound complexity, arise from simpler forms? Hartley posits a law of association by which the simplest sensations aggregate into the most complex ideas. If true, then the minds of humans, dogs, and mites differ only in degree of complexity, not in kind.[1]

[1] Lamarck, by contrast, believed that the mind is endowed with various innate faculties. He attempted, though arguably failed, to square this with evolution by suggesting that new

Hartley's theory of emotion fills a gap in Darwin's theory of evolution. But Darwin's theory also fills a gap in Hartley's. As an empiricist, Hartley believed that the mind begins as a blank slate. A child must learn how to fear, envy, and hope. Yet some emotions, like the love between a mother and her baby, or between a father and his baby, are surely instinctual. They occur spontaneously, without prior experience or learning. Hartley's theory cannot explain instinctual emotions.

To fix Hartley's theory, Darwin appeals to evolution. More specifically, he appeals to heritable habits. A habit in one generation can become an instinct in the next. If mom learns to fear dogs, then baby may be born with an instinctual fear of dogs. (Today, we associate heritable habits with Jean-Baptiste Lamarck's debunked theory of evolution. It is easy – or perhaps convenient – to forget that Darwin also believed in the inheritance of acquired traits and included it within his own theory of evolution.[2])

Darwin's proposal preserves the empiricist thesis that all knowledge and ideas derive from experience, but with the caveat that some knowledge and ideas derive from *one's ancestors'* experiences. In his notes, he defines emotions as "trains of thought long associated with action" (DAR LIB 404, 333). The only difference between this definition and Hartley's is that, for Darwin, "long" can be measured in evolutionary time.

I argue that Darwin's *Expression* is a work of associationist philosophy. It advances an associationist theory of emotion and emotional expression, inspired by Hartley and later associationists. We can see this inspiration most clearly in the three principles of expression, which purport to explain why we express our emotions as we do – why we laugh when amused, frown when angry, and cry when sad.

The first principle ("serviceable associated habits") is purely Hartleian. If a feeling and an action repeatedly co-occur, then an association will form. Subsequently, the feeling will automatically trigger the action.[3] For instance, if angry people tend to get in fights, then they may involuntarily clench their fists when becoming angry, even if they do not intend to fight in this particular instance.

The first principle is based on the principle of contiguity. The second principle ("antithesis") is based on the principle of contrast. Once a feeling

faculties emerge as animals become more complicated. See Gissis (2010) for an analysis of Lamarck's theory of feelings.

[2] I discussed this point in §1.5.1. Compare *Origin* (1859, 134), *Variation* (1868a, II:293–318), *Descent* (1871, I:152–153), and *Expression* (1872, 39).

[3] Hartley was the first person to use the term "automatic" to denote involuntary human behavior (Allen 2001).

becomes associated with an action, the opposite feeling will become associated with the opposite action. Aggressive dogs lower their heads, so affectionate dogs lift them.

The third principle ("the direct action of the nervous system") does not sound like an associationist principle. In positing that excess nerve energy can cause purposeless expressions, independent of habit, it sounds more like a hydraulic principle. However, Darwin inherited this principle from two associationists – Herbert Spencer and Alexander Bain – who developed it in the course of trying to explain the neurophysiology of association. We hear "salt" and think "pepper" because nerve energy travels along the strongest nerve connections. Yet sometimes excess energy is forced across weaker nerve connections.

Taking a step back, we can see that Darwin's *Expression* falls squarely within the associationist tradition. It seeks to explain as much as possible using the principles of association. It embraces Hartley's theory of emotion as a foundation, yet it also seeks to update and revise that theory in light of Darwin's theory of evolution. *Expression* can be read as a postscript to Hartley's *Observations on Man* (1749).[4]

My aim in this chapter is to trace the development of Darwin's theory of emotion and emotional expression, emphasizing Darwin's many and deep engagements with the associationist tradition. At every stage of this development, Darwin was reading associationist philosophy, registering his agreement with it, and articulating responses to objections.

I begin by examining Darwin's notes on expression from 1838–1840. These include Notebook M (1838), Notebook N (1838–1839), the collection of notes that he later titled "Old & Useless Notes" (1838–1840), and the notebook containing his observations of his first child's earliest expressions (1839–1840).[5] During this period, Darwin studied and embraced Hartley's associationist theory of emotion.

For the next twenty-five years, Darwin focused on other topics. He returned to the topic of emotion and expression in the late 1860s. I examine Darwin's "Queries on Expression" (1867) and the research he

[4] Alexander Bain (1873) calls Darwin's *Expression* the postscript to his own *The Senses and the Intellect*. Yet Bain's work is itself a postscript to Hartley's *Observations*.

[5] The notebook of observations of his children's expressions includes notes from later years about his subsequent children, yet these notes become fewer and further between as time passes. I focus on his notes on his first child, recorded at the very end of 1839 through most of 1840. See Conrad (1998) for an analysis of this notebook. Conrad (2004) argues that Darwin's memorial of his daughter Annie, written in 1851, also provides insight into Darwin's early ideas about emotion. I do not examine that memorial text here.

Table 3.1 *Darwin's writings on emotion and emotional expression*

Year(s)	Title	Abbreviation	Published?
1838	Notebook M	M	No
1838–1839	Notebook N	N	No
1838–1840	Old & Useless Notes about the Moral Sense & Some Metaphysical Points	OUN	No
1839–1840	*Notebook of Observations on the Darwin Children*	ODC	No
1867	Queries on Expression	Queries	Yes*
1871	*The Descent of Man*	*Descent*	Yes
1872	*The Expression of the Emotions in Man and Animals*	*Expression*	Yes
1877	A Biographical Sketch of an Infant		Yes

* Darwin distributed an unpublished query in 1867 (Darwin 1867a) and then published the query in 1868 (Darwin 1868b).

conducted leading up to *Descent* (1871) and *Expression* (1872). During this period, Darwin studied the work of the associationists Spencer and Bain. They offered several correctives to Hartley's theory of emotion, and Darwin updated his own theory accordingly.

I aim to show that *Expression* is the culmination of Darwin's philosophical investigations into associationism (see Table 3.1).

3.2 Notebook M (1838)

Darwin returns from his voyage on the *Beagle* in 1836. In the summer of 1837, he opens his first notebook on evolution (Notebook B) and draws a simple phylogenetic tree (B 36). A year later, in 1838, he begins his first notebook on expression (Notebook M). It takes him only two and a half months to fill it from cover to cover.

Howard E. Gruber observes that Notebook M explores a dizzying array of topics without much structure or focus. It jumps from memory to mental illness to free will to beauty (Darwin 1980, 46–49). One reason for this scattershot approach is that Darwin uses M as a reading journal. As he reads articles and books on mind, emotion, and morality, he records his thoughts and impressions in the pages of M.

I shall focus on Darwin's commentary on one particular associationist text, Erasmus Darwin's *Zoonomia* (1796), which inspired the first two principles of expression that would appear thirty years later in *Expression*.[6]

3.2.1 *Erasmus Darwin's* Zoonomia

In 1838, Charles embarks on a close reading of Erasmus Darwin's *Zoonomia* (1796). Erasmus was Charles's grandfather, his father's father, who died before Charles was born. Erasmus was a towering figure in the Midlands Enlightenment. He wrote two scientific poems, published together as *The Botanic Garden* (1791), which introduced general audiences to the Linnaean classification of plants.[7] He also wrote a lengthy prose work, the *Zoonomia; or The Laws of Organic Life* (1796), which is essentially a textbook on biology, psychology, and medicine (Montgomery 1985, 39). Today, the *Zoonomia* is remembered for its suggestion of evolution, based on the inheritance of acquired characteristics, which predates Lamarck's similar theory.[8] Charles would later deny that Erasmus's theory of evolution influenced his own. Yet when Charles begins formulating his own theory of evolution in 1837, he titles his first notebook on the subject "Zoonomia" (2009, B 1).

Erasmus Darwin founded an intellectual society called the Lunar Society of Birmingham. One of its members, the philosopher-chemist Joseph Priestley, was an apostle for David Hartley's associationist philosophy. Priestley published a popular abridgment of Hartley's *Observations on Man* (1749), retitled *Hartley's Theory of the Human Mind* (1775). And he later wrote that Hartley's work was "without exception, the most valuable production of the mind of man" (Priestley 1782, II:161; quoted in Cairns 2007, 41). Priestley's passion for Hartley clearly rubbed off on Erasmus Darwin, who presents and defends Hartley's associationist theory of emotion in the psychology sections of the *Zoonomia*.[9]

Echoing Hartley's claim that "all the Passions arise thus from Pleasure and Pain" (1749, prop. 89), Erasmus writes, "All our emotions and passions seem to arise out of the exertions of these two faculties of the

[6] Parts of the following sections (§§3.2.1–3.2.4) were previously published in Glazer (2024).
[7] See Porter (2018a; 2018b) for insightful analyses of Erasmus's poetry.
[8] King-Hele (1999, 297–301) provides an accessible reconstruction of Erasmus's evolutionary theory. Schofield is somewhat dismissive of the claim that Erasmus had an evolutionary theory: "Darwin was more interested in developing a theory of epigenesis to support his medical hypothesis than he was in propounding a mechanism of evolution" (1963, 396).
[9] Smith writes of the *Zoonomia* that "Hartley also glimmers through Darwin's associationist psychology" (2005, 140). See also Smith's (1987) account of Hartley's associationist psychology.

animal sensorium" (1796, 11.2.2). Echoing Hartley's claim that anger is a train of associations connecting the feeling of hurt with the desire to hurt (1749, prop. 97), Erasmus writes that "anger includes the pain from a recent injury, and the aversion to the adversary that occasioned it" (1796, 11.2.2). Finally, echoing Hartley's claim that mental associations are caused by nerve vibrations, Erasmus writes that every mental association results from "an exertion or change of . . . the sensorium" (1796, 5.2). Erasmus mentions Hartley only once by name in the *Zoonomia* (1796, 39.1), but its sections on psychology are unmistakably Hartleian.[10]

Here is why these details matter. When Charles Darwin first begins theorizing about emotion and emotional expression in 1838, he reads and takes careful notes on his grandfather's presentation of Hartley's associationist theory of emotion. He copies two principles from the *Zoonomia*, which appear thirty-five years later, almost entirely unchanged, as his first two principles of expression in *Expression* (1872). We can draw straight lines from Erasmus's *Zoonomia* (1796), through Charles's Notebook M (1838), to Charles's *Expression* (1872).

3.2.2 The Principle of Serviceable Associated Habits

Inspiring Charles's first principle of expression, Erasmus theorizes that expressions are movements that have become strongly associated with particular sensations, such that the sensations automatically elicit the movements. For instance, he hypothesizes that the expressions of fear are based on associations formed in the traumatic event of birth:

> As soon as the young animal is born, the first important sensations, that occur to him, are occasioned by the . . . want of respiration, and by his sudden transition from ninety-eight degrees of heat into so cold a climate. – He trembles . . . and begins to breathe with frequent and short respirations; at the same time the cold contracts his red skin, gradually turning it pale; the contents of the bladder and of the bowels are evacuated.. . . This early association of motions and sensations persists throughout life; the passion of fear produces a cold and pale skin, with tremblings, quick respiration, and an evacuation of the bladder and bowels, and thus constitutes the natural or universal language of this passion. (1796, 16.8.1)

In the first moments of life, trembling and gasping for breath are serviceable. They help the newborn to adapt to the new environment

[10] See Simili (2005, 157), Uglow (2005, 174), Valsania (2005, 342), and Fara (2012, 239) for brief acknowledgments of Hartley's influence on Erasmus's associationist psychology.

outside the womb. Based on this single traumatic event, a habit forms. Even decades later, once the infant becomes an adult, they continue to tremble and gasp when afraid, even though these actions no longer serve any purpose.

To be clear, the claim is not that the infant forms an association between *fear* and trembling, since this would imply that fear exists prior to the association. Rather, the claim is that the infant forms an association between *a series of sensations* and trembling. Fear doesn't occur until afterward, when one of those sensations sets in motion a train of associations, including copies of those other sensations (now "ideas" and "feelings") and the act of trembling. On Hartley's theory, which Erasmus adopts, emotions are the products of association, sequences of sensations and movements.

Erasmus provides further examples of this first principle: "The actions that constitute the mode of fighting, are the immediate language of anger in all animals; and a preparation for these actions is the natural language of threatening" (1796, 16.8.5). And: "when we hear any the smallest sound, that we cannot immediately account for, our fears are alarmed, we suspend our steps, hold every muscle still, open our mouths a little, erect our ears, and listen to gain further information" (1796, 16.8.6).

Charles Darwin copies this last example in Notebook M: "Fear is connected with *habitual* stopping of breath to hear any sound" (2009, M 131e). He rewords it only slightly for *Expression*: "The diversified expressions of fear . . . may be accounted for through the principles of habit, association . . . to hear distinctly whatever sound may reach our ears" (1872, 307). In both M and *Expression*, Charles includes Erasmus's observation that habitual expressions will still occur, even when they serve no purpose. He writes, "Fear is open mouthed to *hear*, though in individual case, nothing can be heard" (Darwin 2009, M 149), and that "there is a tendency through the force of habit and association for the same movements to be performed, though they may not then be of the least use" (1872, 28).

In sum, Darwin's first principle of expression, the principle of serviceable associated habits, is derived from Erasmus Darwin's Hartleian theory of emotion, which Darwin studied in 1838.

3.2.3 *The Principle of Antithesis*

Inspiring Charles's second principle of expression, Erasmus hypothesizes that some expressions are the *contrasting* movements that reflexively occur after contrasting sensations. He observes that when an infant's hunger is sated, "the antagonist muscles of the face gently acting, produce the smile of pleasure" (1796, 16.8.4). Pained by hunger, the infant puckers its lips to latch to the breast. When hunger passes over into satisfaction, the lips move in an opposing manner, stretching into a broad smile. Smiling serves no purpose. It arises in opposition to the serviceable movements that accompany opposing sensations.

Notice how similar Erasmus's analysis is to Charles's principle of antithesis from *Expression*: "Now when a directly opposite state of mind is induced, there is a strong and involuntary tendency to the performance of movements of a directly opposite nature, although these are of no use" (1872, 28). This is no coincidence. The second principle, like the first, is copied directly from Erasmus's *Zoonomia* into Charles's Notebook M.[11]

Compare the following two passages, the first from Erasmus's *Zoonomia*, and the second from Charles's Notebook M:

> Hence this smile during our lives is associated with gentle pleasure; it is visible in kittens, and puppies, when they are played with, and tickled; but more particularly marks the human features. For in children this expression of pleasure is much encouraged, by their imitation of their parents, or friends; who generally address them with a smiling countenance. (E. Darwin 1796, 16.8.4)

> Why does dog put down ears, when pleased. – is it opposite movement to drawing them close on head, when going to fight, in which case expression resembles a fox – I can conceive the opposite muscles would act, to when in a passion. – dog tail curled when angry & very stiff, back arched, just contrary, when pleased tail loose & wagging . . . it is very curious, recurrence of pleasure so teaching expression as constant smiles, cheerful face. – Man when at ease has smooth brow contrary to wrinkled.. . . Cat when pleased, erect its tail & make it very stiff & back when savage no & ready to dash at prey streched [*sic*] out & flaccid, when furious with fright back absurdly arched. & tail stiff. (C. Darwin 2009, M 146e–147)

Both passages mention cats, dogs, and parents teaching/encouraging children to smile. I have little doubt that Charles wrote the second passage after reading the first. However, while Erasmus focuses narrowly on smiling as an opposing expression, Charles extends the principle to account for other opposing expressions.

11 Montgomery argues that Charles's first principle is derived from Erasmus's theory but that "there is nothing comparable to [Charles's second principle] in Erasmus's book" (1985, 41). I disagree.

In sum, Charles Darwin did not come up with either of his first two principles of expression. He inherited both from his grandfather, Erasmus Darwin, who borrowed the first, if not the second, from David Hartley.

3.2.4 *Charles Darwin's Evolutionary Twist*

I do not mean to suggest that Charles's early theorizing about emotion and expression is entirely unoriginal. To the contrary, Charles identifies a problem with Erasmus's theory and proposes a clever solution.

The problem is that some emotions appear to be instinctual – that some trains of associations occur prior to the experiences needed to learn them. Darwin puts the problem like this: "Starting must be habitual involuntary movement from wish to avoid some danger – but it is instinctive because Nancy [Darwin's nurse] tells me very young babies start at anything they hear or see which frightens them" (2009, M 52). Erasmus, following Hartley, claims that babies must learn to fear, yet Charles believes that some fears are innate.

Rather than rejecting Erasmus's theory, Charles proposes an amendment: Emotions are indeed the products of association, as Erasmus claims, but associations can be inherited, presenting in subsequent generations as innate instincts. Charles writes:

> If one performs some actions, which are pleasant, every concomitant circumstance calls up pleasure, or pleasure or pain of association. – now if one has these feelings, without being aware of their association ie heredetary [*sic*], does one not call them instinctive emotions? (2009, M 127).

More directly: "Expression, is an heredetary [*sic*] habitual movement consequent on some action, which the progenitor did, when excited or disturbed by the same cause, which now excites the expression" (2009, M 107).

In other words, Darwin proposes a synthesis of Erasmus's associationist theory of emotion with his own budding theory of evolution (see Dixon 2003, 162–163). All emotions derive from experience, but not all emotions derive from one's own experience; some emotions derive from one's ancestors' experience (C. Darwin 2009, M 128).

Keep in mind that Darwin has not yet discovered natural selection, although he will discover it within a matter of weeks (Montgomery 1985, 39). By this point in September of 1838, he is convinced *that* species evolve, but his view of *how* species evolve is still thoroughly "Lamarckian." Environments change, individuals learn to adapt, and their offspring inherit these adaptations.

3.3 Notebook N (1838–1839)

It took Charles only two and a half months to fill Notebook M. When he ran out of space in M, his musings on expression spilled over into Notebook N. The most important sections of Notebook N, for our purposes, are those in which he engages with Hartley's theory of emotion.[12]

Prior to 1839, Hartley's influence on Charles Darwin was indirect. Charles was familiar with his grandfather's theory, but not with the theory that inspired it. This changed in 1839 when Charles read Sir James Mackintosh's *Dissertation on the Progress of Ethical Philosophy* (1837).[13] Originally part of the *Encyclopædia Britannica*, Mackintosh's *Dissertation* contains summaries and evaluations of the great British philosophers, including Hartley. Charles took careful notes on the chapter on Hartley. It is in these notes that we see Charles delving into the nature of emotion.[14]

Recall that on Hartley's view, emotions are trains of associated thoughts, feelings, and actions. Fear is a sequence of painful associations hurtling us toward fight, flight, or freeze. Mackintosh rejects this view in his critical commentary (Manier 1978, 98). He claims that emotions are just feelings. "Primary" emotions arise spontaneously in response to events, without any association involved. "Secondary emotions" arise from associations among the primary emotions and other thoughts.[15] However, they too are experienced as singular feelings. In Hartley's view, the process of association combines elements into decomposable aggregates. But in Mackintosh's view, the process of association fuses emotions and ideas into new, indecomposable feelings (1837, 254–255).

Darwin reads Mackintosh's summary and critique of Hartley carefully and then remarks in Notebook N: "Mackintosh first clearly insisted on assoc[iation] of ideas & emotions. Rather ideas & bodily actions make the

12 Parts of this section (§3.3) were previously published in Glazer (2024).

13 Darwin was personally acquainted with Mackintosh. In his autobiography, Darwin wrote that Mackintosh was "the best converser on grave subjects to whom I ever listened" (1958, 66). Another philosopher, William Whewell, wrote the preface to Mackintosh's *Dissertation*. Darwin was also personally acquainted with Whewell and called him the second-best converser on grave subjects.

14 Dixon (2003) discusses Darwin's engagement with Mackintosh but overlooks Darwin's theory of emotion. In Dixon's assessment, "[Darwin] did not have any theory of emotions" (2003, 166). Richards (2009, 101–102) also discusses Darwin's reading of Mackintosh but focuses on the latter's theory of the moral sense.

15 Notice the similarity between Mackintosh's distinction between primary and secondary emotions and Hume's distinction between direct and indirect passions, which I summarized in §2.4.

emotions" (2009, N 89). In other words, Darwin takes Hartley's side in the debate.[16] An emotion is not a raw material of association. Rather, an emotion is a train of "ideas & bodily actions," a product of association. Darwin repeats this point in the margins of his personal copy of Mackintosh's *Dissertation*. Beneath the claim that emotions exist prior to association, Darwin writes, "yet emotions are results – are trains of thought long associated with action" (DAR LIB 404, 333). The wording is slightly amphibolous. He might be saying that emotions are trains of thought, which may then become associated with action. This reading implies that the emotion is one thing and the expression another. Alternatively, he might be saying that emotions are trains of thought and action, which have become associated. This reading implies that the expression is part of the emotion.

One reason to think he intends the latter is that he explicitly considers its logical consequence: If the expression is part of the emotion, then, strictly speaking, the emotion is not the cause of the expression.[17] Darwin asks: "but what first caused this bodily action, if the emotion was not first felt?" (2009, N, 66). He answers: "without abrupt flush, acceleration of pulse, or rigidity of muscles, – man cannot be said to be *angry*. He may have pain or pleasure these are sensations" (2009, N, 66). When we are angry, we experience familiar feelings. However, these feelings are what Hartley calls the "nascent" state of the emotion. They are the initial links in the chain of associated feelings and actions that constitutes anger.

An important caveat is needed here. The view that expressions are parts of emotions implies that emotions are not the causes of expressions. However, it is perfectly consistent with the commonsense view that emotional *feelings* are the causes of emotional *expressions*. An emotion is a train of associations, each associate causing the next. A sensation causes a feeling, which causes another and then another. But at some point, the feelings start causing bodily movements. Thus, we can coherently say that when a person experiences, say, the emotion of anguish, their suffering causes them to cry. The feeling causes the expression and the whole thing is the emotion. Throughout *Expression*, Darwin writes that expressions occur "under the influence" of emotions (e.g., 1872, 27, 65, 86, 100, 192, 242, 355). This language certainly suggests that the emotion *is* the cause of expression, yet Dewey argued early on that this language serves "only to

16 Manier (1978, 99–100) explores Darwin's rejection of other aspects of Mackintosh's philosophy.

17 Dewey (1894) and Campbell (1997a) both emphasize this consequence of Darwin's theory of expression.

confuse" (1894, 556), since Darwin rejects the commonsense view that a "pre-existent emotion" is the cause of the expression (1894, 555).[18]

Common sense says that there is no expression without an emotion. Darwin adds that there is no emotion without an expression either (Manier 1978, 136). In the margins of his copy of Mackintosh, he writes, "Emotions, having been formed by actions, will always lead to them" (DAR LIB 404, 381; quoted in Dixon 2003, 166). There may be a train of associated thoughts and feelings. But this train only becomes an *emotion*, an experience that *moves* us, once the train includes physical movements as well. The obvious objection is that we sometimes hide our emotions. We feel anger or sadness or fear, but we keep a straight face. Darwin does not countenance this objection in Notebook N, but he addresses it decades later in *Expression* (1872, 28–29), arguing that willful attempts to inhibit expressions inevitably lead to other unintended and involuntary expressions. Though others may overlook or misunderstand these alternative expressions, the fact remains that emotions are always expressed.

If trains of thoughts and feelings do not become emotions until expressions are connected to the end, then a theory of the origins of expressions is *eo ipso* a theory of the origins of emotions (Darwin 2009, N, 89). The moment at which thoughts of injury and feelings of unease become linked to the tendency to flee is the moment at which fear is born.

3.4 "Old and Useless Notes" (1838–1840)

Darwin continued to develop his theory of emotion in another set of notes, which he later compiled under the doubly misleading title, "Old & useless Notes about the moral sense & some metaphysical points written about the year 1837 & earlier" (henceforth OUN). First, Darwin got the dates wrong. These notes were in fact written between 1838 and 1840. The notes we will examine were written in 1839. Second, Darwin gave the collection this title while preparing his *Descent of Man* (1871) and called them "useless" only because he did not include them in that work.[19] These notes became useful again when he wrote *Expression* (1872).[20]

[18] George Herbert Mead (1967, 17) attributes to Darwin the view that emotions are mental states that cause their behavioral expressions. I review Dewey's (1894) argument for why, appearances notwithstanding, Darwin's view is that expressions are parts of emotions in §6.3. Neither Mead nor Dewey, however, considers Darwin's debt to associationism.

[19] Sandra Herbert and Paul Barrett offer these two corrections in their introduction to this collection of notes (Darwin 2009, 597).

[20] Parts of this section were previously published in Glazer (2024).

So far, we have the idea that emotions are trains of thought and action. Hartley, like Erasmus, claims that these trains derive from associations formed in the course of a lifetime. Charles adds his trademarked evolutionary twist: If some associations were formed in one's ancestors' lifetimes, then the resulting emotion would be instinctual.[21] Charles works out this view in his commentary on Mackintosh's summary of Hartley:

> Hartley . . . explains our love of another, as pleasure arising from association from having received benefits from this person. – But the love is instinctive, & how does it apply to mother loving child, from whom, she has never received any benefit. – Yet I think there is much truth in doctrine (2009, OUN 49).

Darwin identifies a gap in Hartley's theory: Some emotions occur prior to their supposedly formative experiences. Yet Darwin is hopeful that this gap can be filled. He then asks:

> But does not Hartley explanation apply perfectly to origin of these instincts. – the having received pleasure from some *one* person in early infancy, during many generations giving love of mother; the having received some advantages from man. during many generations giving the social feelings. (2009, OUN 50)

In other words, there is only a problem with Hartley's account if we insist that emotions derive from one's own experiences. Allowing for emotions to derive from one's ancestors' experiences solves the problem. A mother's instinctive love for her baby may be the result of many generations of mothers developing love for their children.

Emotions are trains of thought and action, linked by association. *Some* of these associations are inherited, but Darwin insists that new associations are learned, and old associations are unlearned, in the course of experience. Emotions are not hardwired. They are preprogrammed, to an extent, but they are reprogrammed repeatedly throughout one's life. He compares emotions to tastes, claiming that each is a "hereditary habit (& modified & associated during lifetime)" (Darwin 2009, OUN 50^{v}). Evolution plays a role in shaping our emotions, but so does experience. Both operate according to the same underlying principle: association.

Although Darwin acknowledges the importance of evolution and experience, he clearly values the contributions of evolution more. In the

21 Notice the parallel with his analysis of expression. In both analyses, associationism provides the foundation, and evolution enters the picture later to explain how some associations may precede experience.

case of moral emotions, he cautions that the cultural reprogramming of emotion is often at humankind's expense.[22] He says that the "part which is acquired by association from education & imitation, has often been perverted from want of reason" (Darwin 2009, OUN 51). Inherited moral emotions: good. Learned moral emotions: bad. Furthermore, "I should think some parts of the emotive part of man, may be quite artificial, as avarice love of gold. – love of fame – Yes Hartley explains this" (Darwin 2009, OUN 54). Nature doesn't make us bad. Nurture is to blame. And there's Hartley again!

Note that Darwin has not abandoned his "Lamarckian" theory of emotion, even though he has, by this point in 1839, identified natural selection as an alternative mechanism by which species evolve.

Taking a big step back, we can see that Darwin first developed an associationist theory of emotion and emotional expression, inspired by Hartley and Erasmus, between 1838 and 1839. He defined emotions as "trains of thought long associated with action" (DAR LIB 404, 333). And he formulated two principles of association to explain why we express our emotions as we do. These two principles formed the foundation of *Expression*, published more than thirty years later.

3.5 *Notebook on Observations on the Darwin Children* (1839–1840)

When Darwin's first child was born in December of 1839 – eight years to the day after starting his *Beagle* voyage – his interests turned from the theoretical to the empirical. He began filling another notebook, now titled *Notebook on Observations on the Darwin Children* (henceforth ODC), with observations of his newborn son's expressions. Many years later, he polished and published these notes as "A Biographical Sketch of an Infant" (1877) in the philosophy journal *Mind*.[23]

In both the original notebook and the published article, Darwin focuses on identifying instinctual expressions – those that cannot have been learned. Since Darwin considered evolution, and more specifically heritable habits, to be his own unique contribution to the associationist theory

[22] Darwin was particularly interested in the moral emotions. In 1839 he began a manuscript on the moral sense but never completed it. However, he returned to this topic in *Descent*. Famously, he argued that our sympathy for the poor and weak prevents us from allowing natural selection to run its course (Rogers 1972, 273–274; Richards 2009, 111–113).

[23] See Lorch & Hellal (2010) and Conrad (1998) for analyses of Darwin's study of babies.

of emotion, it makes sense that he would look for empirical evidence of instinctual emotions.[24]

In Notebook M, he records his nurse Nancy's claim that infants startle in surprise almost immediately after birth (2009, M, 52).[25] Following the birth of his son, he is able to confirm this fact for himself: "The Baby started certainly during first fortnight at sudden sounds" (ODC, 2v). Darwin later adds that the Baby reliably blinks when startled, even when the cause is a loud sound. He reasons: "He never could have learnt from experience to close his eyes very suddenly from sudden noise near his head. – . If it had been experience he would have done it, from visible object – This blinking of eyes, I dont doubt instinct to protect eyes" (ODC, 13). Darwin repeats this point in *Expression* (1872, 38–39).

Darwin notes that William frowned in his first week of life, inferring that this expression too is instinctual. He notes that his son's vision is "exceedingly imperfect" – indeed, it is not until the child's ninth day that he appears to be able to follow the flame of a candle with his eyes (ODC, 2). Darwin then reasons that if the function of frowning is to assist vision, and if babies frown before being able to see properly, then frowning must be instinctual (ODC, 2).

Darwin does not observe smiling until William is around six weeks old.[26] He speculates that smiling occurs now that his son's vision has developed enough to recognize faces. One day he writes, "Emma argues that his smiles were from seeing her face" (ODC, 3Bv). Although he seems somewhat doubtful of Emma's conclusion here, he is clearly convinced by the next day, when he writes that his son "smiled repeatedly, & I think chiefly when suddenly seeing face, of mother. & mine" (ODC, 4). Darwin's inference is that smiling with joy is instinctual, yet a child must first develop the ability to perceive the stimuli that elicit joy. Darwin later notes the week that his son appeared to take pleasure in the sound of music (ODC, 14).

Darwin does not assume that all of his son's expressions are instinctual. He notes multiple examples of learned expressions, based on acquired habits (ODC, 13) and imitation (ODC, 15).

24 Darwin claimed to arrive at his principles only after a careful consideration of the facts, yet the truth is that Darwin came up with the principles first and then conscientiously investigated the facts afterward. The same is true of his theory of natural selection. He first formulated the logic of natural selection in 1838, but then spent twenty years assembling the mountain of facts that support his theory.

25 Browne (2002, 305; 1985) points out that Darwin viewed mothers and nurses as underappreciated experts on infant behavior.

26 Darwin observed a smile at nearly five weeks of age but was sure that it was "a chance movement of muscles, without a corresponding sensation" (ODC, 3).

Darwin was not always a passive observer. He conducted amusing little experiments on his son to see how he would respond. For instance:

> 4 1/2 months old. I made loud snoring noise, near his face, which made him look grave & afraid & then suddenly burst out crying. This is curious, considering the wondrous number of strange noises, & stranger grimaces I have made at him, & which he has always taken as good joke. I repeated the experiment. (ODC, 17)

A similar story appears in *Expression*:

> When this child was about four months old, I made in his presence many odd noises and strange grimaces, and tried to look savage; but the noises, if not too loud, as well as the grimaces, were all taken as good jokes; and I attributed this at the time to their being preceded or accompanied by smiles. (1872, 359)

The moral of the story in ODC is that the child expressed fear. The moral of the story in *Expression* is that the child must have understood Darwin's smiles.

Darwin was interested not only in his son's expressions, but also in his son's ability to form new associations. He notes: "During last week has shown considerable association of ideas. – He understands, when his cloak is on, he is to go out & is cross, if not taken immediately" (ODC, 19). Again: "He now understands a good many things, & will associate almost any sentence with some object if repeated two or three times at interval" (ODC, 25). Finally: "Nothing has struck me more in his intellectual development, compared with puppy, that the great quickness of associating any two things together, after they have happened even only twice or thrice together" (ODC, 27).

Between Darwin's detailed examples and amusing anecdotes, we can sense the underlying associationist theory of the mind. A child is born with some inherited associations but soon begins forming many more.

3.6 The Twenty-Five-Year Gap (1841–1866)

For the next twenty-five years, Darwin focuses on other projects. He writes mostly about geology in the 1840s, barnacles in the 1850s, and plants in the 1860s. During this period, Darwin writes virtually nothing on emotion or emotional expression. He discusses his deceased daughter Annie's emotions in a memorial text composed in 1851 (Conrad 2004). And in a letter to Herbert Spencer from 1858 – the year before *Origin* was published – Darwin notes that "Expression has been for years a favourite subject with me for *loose* speculation, & I most entirely agree with you that all expression has some biological meaning" (Darwin 1858). Yet Darwin appears not to have put any more speculation in writing.

3.7 Queries about Expression (1867)

Darwin finally returns to the topic around 1867, as he begins working on an essay on the evolution of humankind, which eventually grows and splits into *Descent* (1871) and *Expression* (1872). Despite the time that had elapsed, Darwin picks up right where he left off: the careful pursuit of facts to support the theory he had developed in 1838–1839.

In 1867, he sends a questionnaire to recipients around the globe, asking them how the locals express their emotions.[27] Darwin's questionnaire contained seventeen questions:

1. Is astonishment expressed by the eyes and mouth being opened wide, and by the eyebrows being raised?
2. Does shame excite a blush when the colour of the skin allows it to be visible? and especially how low down the body does the blush extend?
3. When a man is indignant or defiant does he frown, hold his body and head erect, square his shoulders and clench his fists?
4. When considering deeply on any subject, or trying to understand any puzzle, does he frown, or wrinkle the skin beneath the lower eyelids?
5. When in low spirits, are the corners of the mouth depressed, and the inner corner of the eyebrows raised by that muscle which the French call the "Grief muscle"? The eyebrow in this state becomes slightly oblique, with a little swelling at the inner end; and the forehead is transversly wrinkled in the middle part, but not across the whole breadth, as when the eyebrows are raised in surprise.
6. When in good spirits do the eyes sparkle, with the skin a little wrinkled round and under them, and with the mouth a little drawn back at the corners?
7. When a man sneers or snarls at another, is the corner of the upper lip over the canine or eye tooth raised on the side facing the man whom he addresses?

[27] The Darwin Correspondence Project at the University of Cambridge has created an interactive map, which allows users to find Darwin's recipients. Visit www.darwinproject.ac.uk/world-map/Expression%20questionnaire.

8. Can a dogged or obstinate expression be recognized, which is chiefly shewn by the mouth being firmly closed, a lowering brow; and a slight frown?
9. Is contempt expressed by a slight protrusion of the lips and by turning up the nose, with a slight expiration?
10. Is disgust shewn by the lower lip being turned down, the upper lip slightly raised, with a sudden expiration, something like incipient vomiting, or like something spat out of the mouth?
11. Is extreme fear expressed in the same general manner as with Europeans?
12. Is laughter ever carried to such an extreme as to bring tears into the eyes?
13. When a man wishes to shew that he cannot prevent something being done, or cannot himself do something, does he shrug his shoulders, turn inwards his elbows, extend outwards his hands and open the palms; with the eyebrows raised?
14. Do the children when sulky, pout or greatly protrude the lips?
15. Can guilty, or sly, or jealous expressions be recognized? though I know not how these can be defined.
16. As a sign to keep silent, is a gentle hiss uttered?
17. Is the head nodded vertically in affirmation, and shaken laterally in negation? (Darwin 1867a; 1868b)

Darwin reproduced this query in *Expression* (1872, 16), but without question 16, the only item dealing with the vocal expression of emotion.[28]

Darwin reports that he received thirty-six replies, which convinced him that "the same state of mind is expressed throughout the world with remarkable uniformity" (Darwin 1872, 17). One could object that the replies do not actually support Darwin's conclusion. Alfred Russell Wallace's reply, for example, highlights multiple differences between European and Malay expressions:

> For the Malays I should say Yes. to 1. 3. 8. 9. 10. & 17. and No. to 12. 13. and 16. but I cannot be certain in any one. But do you think these things are of much importance? I am inclined to think that if you could get good direct observations you would find some of them often differ from tribe to tribe, from island to island and sometimes from village to village. (Wallace 1867)

[28] See Freeman and Gautrey (1972; 1975) for a critical examination of the various versions of the query that exist.

Similarly, Mary Elizabeth Barber's reply mentions several ways in which indigenous South African expressions differ from European expressions:

> Answers to some of the queries about expression in the Native races of S. Africa . . .
> No. 3. I have never seen a Kafir or Fingoe clench his fists, they do no fight with their fists.
> No. 11. Yes, in very much the same way as in Europeans. I have often heard it said of Kafirs and Fingoes that they were pale with rage or fear.
> No. 12 – Yes. they frequently laugh until the tears run down their cheeks, especially the women.
> No. 13. I have never seen a Kafir or Fingoe shrug his shoulders or extend outwardly the pals of his hand. (Barber 1867)

Yet Darwin does not shy away from mentioning these cultural differences in *Expression.* For instance, he emphasizes the cultural relativity of clenching fists in anger, fully agreeing with Barber's reasoning that people who do not fight with their fists will not clench their fists in anger either:

> Rage, anger, and indignation are exhibited in nearly the same manner throughout the world.... There is, however, an exception with respect to clenching the fists, which seems confined chiefly to the men who fight with their fists. With the Australians only one of my informants has seen the fists clenched.... Another observer speaks of the native men, when enraged, throwing their arms wildly about. (Darwin 1872, 246–247)

Darwin likewise mentions that while the shrugging of shoulders has been observed in several cultures, "Mrs. Barber has never seen this gesture with the Kafirs of South Africa" (1872, 269).

Recall that in Darwin's 1838–1839 theory, both evolution and experience influence our emotions, since both operate on the same underlying principle, that of association. Sensations set in motion trains of associations. Some of these associations are inherited. Others are learned. Emotions are not hardwired and fixed. They are preprogrammed, to an extent, but they can be and often are reprogrammed in our lifetimes.

Darwin in fact drew two different conclusions from the responses to his queries. The first conclusion is that there are inherited, universal expressions of emotion. The second conclusion is that there are also learned, culturally specific expressions of emotion. The latter are, however, deviations from the norm.

Those who focus on Darwin's claim of universality overlook an important line of argumentation throughout *Expression*, namely, the role that education plays in the transformation of our expressions. On this point, Darwin's views are regrettably prejudicial. He suggests that we can distinguish learned from inherited expressions by contrasting the expressions of the "sane" with the "insane," of the "civilized" with the "uncivilized," and of adults with children. For instance: "Nor is it an anomalous fact that the children of savages should exhibit a stronger tendency to protrude their lips, when sulky, than the children of civilized Europeans; for the essence of savagery seems to consist in the retention of a primordial condition" (Darwin 1872, 234–235). The "children of savages" perform the inherited expression, while the "children of civilized Europeans" are in the process of unlearning that expression. Darwin questioned some of the racist and colonialist views of his day, but he accepted many others.

The reason why Darwin includes so many photographs of young children in *Expression* is that he believes that they embody the pure state of natural inheritance, unaffected or only marginally affected by social learning. These photographs stand in stark contrast to those of the adults Oscar Rejlander, his wife Mary, and a young man, whose expressions are clearly pantomimed.

3.8 Research for *Descent* and *Expression* (1867–1872)

Sometime around 1867, Darwin's interests shifted back from the empirical to the theoretical. He began reading associationist philosophy again, especially the works of Herbert Spencer and Alexander Bain. Spencer inspired Darwin to add a third principle of expression ("the direct action of the nervous system"). Bain inspired Darwin to supplement his first principle ("serviceable associated habits").

3.8.1 Herbert Spencer

Darwin's private letters from the early 1860s reveal his reluctance to read Spencer's work.[29] However, Spencer presents an associationist theory of emotion in *Principles of Psychology* (1855) and "The Physiology of Laughter" (1860).[30] Darwin did apparently read these works by the end of the decade, given that he quotes from them in both *Descent* (1871) and *Expression* (1872).[31]

29 See, for instance, Darwin's letter to J. D. Hooker on June 23, 1863 (Darwin 1863), Darwin's letter to A. R. Wallace on or around April 10, 1864 (Darwin 1864), and Darwin's letter to Charles Lyell on March 25, 1865 (Darwin 1865).

30 Parts of this section were previously published in Glazer (2024).

31 Darwin's personal library includes two dozen books by Spencer, including the first and second editions of the *Principles of Psychology* and several volumes of Spencer's *Essays*.

Herbert Spencer's father was a member of the Derby Philosophical Society, another club founded by Erasmus Darwin. Young Herbert studied Erasmus's *Zoonomia* (1796) and Lamarck's *Zoological Philosophy* (1984), which present similar theories of evolution. These early experiences were formative, as Spencer would spend the rest of his life developing and applying his own theory of evolution.

In the *Principles of Psychology* (1855), Spencer, like Erasmus, adopts Hartley's theory of emotion. Fear, he writes, is having "in a slight degree those psychical states accompanying the reception of wounds, those which express themselves in cries, those which are experienced during flight" (1855, §200). In other words, fear is a train of associations including the thoughts, feelings, and actions that previously occurred in the context of facing danger. These components are not intrinsically related to each other. They are extrinsically related to each other through the power of association.

Spencer then identifies the same problem with associationist theories of emotion that Charles Darwin had identified in his unpublished notes from 1838–1839: "The doctrine maintained by some philosophers, that all the desires, all the sentiments, are generated by the experiences of the individual, is so glaringly at variance with hosts of facts, that I cannot but wonder how any one should ever have entertained it" (1855, §203). It is obvious, Spencer claims, that some emotions are instinctual. He proposes the same solution that Darwin had proposed: The associations constituting these instinctual emotions have been inherited (1855, §203). One generation's a posteriori associations become the next generation's a priori associations.

Spencer is no mere follower. Drawing on contemporaneous advances in neurophysiology, Spencer replaces Hartley's antiquated theory of nerve vibration with a newfangled theory of nerve force. Long story short, Spencer claims that trains of associations are caused by the flow of nerve force along nerve fibers, which are strengthened or weakened based on experience. When two ideas occur together, like "salt" and "pepper," the connection between nerves is strengthened. When two ideas occur apart, like "salt" and "leper," the connection between nerves is weakened. Spencer's theory presages Hebb's law: *Neurons that fire together wire together.*

But sometimes, the flow of nerve force hits a bottleneck. When this happens, "an overflow of nerve-force undirected by any motive, will manifestly take first the most habitual routes; and if these do not suffice, will next overflow into the less habitual ones" (Spencer 1860, 398). In other words, a sensation will set in motion the normal train of associations, but something else will happen, too. The excess nerve force will activate nerves that are only weakly connected, eliciting unexpected thoughts, feelings, and actions. Famously, Spencer (1860) applies the principle of overflow to explain laughter (Morreall 2023). Someone asks you, "Why did the chicken cross the road?" Your gears start turning. You think about what the chicken might want. You think about why the chicken is by the road in the first place. When you hear the answer, "to get to the other side," you realize that all the pondering was for naught. You realize that you took the question much too seriously. Spencer hypothesizes that your nervous system generated nerve energy to move the wheels of thought. But the wheels of thought screeched to a halt when you heard the punch line. The leftover nerve energy needed to go somewhere. Spencer argues that it spills over into the respiratory system, leading to an abrupt voiced exhale: "Ha!" (Once you've heard the joke a few times, you know what to expect, and so your nervous system stops generating nerve force to turn the wheels of thought.)

Darwin begins *Expression* by criticizing older theories of expression. "The older theories [on physiognomy], which I have consulted, have been of little or no service to me" (1872, 1). He praises anatomists, including Sir Charles Bell and Guillaume Duchenne, for correctly describing the muscles involved in many expressions, but he laments that they do not attempt to explain why we express our emotions as we do (1872, 3, 6). When Darwin turns to Spencer's theory, however, criticism gives way to adulation:

> Mr. Herbert Spencer, in treating of the Feelings in his "Principles of Psychology" (1855), makes the following remarks: "Fear, when strong, expresses itself in cries, in efforts to hide or escape, in palpitations and tremblings; and these are just the manifestations that would accompany an actual experience of the evil feared. The destructive passions are shown in a general tension of the muscular system, in gnashing of the teeth and protrusion of the claws, in dilated eyes and nostrils, in growls; and these are weaker forms of the actions that accompany the killing of prey." Here we have, as I believe, the true theory of a large number of expressions; but the chief interest and difficulty of the subject lies in following out the wonderfully complex results. (1872, 9)

Why do we express fear with crying and trembling? Because crying and trembling "accompany the actual experience of the evil feared." Based on a traumatic experience, in which various sensations, feelings, thoughts, and actions co-occur, associations form, such that the mere thought of danger can set in motion a train of its associates, ending in crying and trembling. We call this train of associations "fear," and we say that "Fear . . . expresses itself in cries ... and trembling." But really it is the power of association doing all the work. Emotions are the products of association, the resulting aggregates of elementary thoughts, feelings, and actions. Darwin calls Spencer's theory "the true theory of a large number of expressions," but it is not merely a theory of expression; it is a robust theory of emotion. If Spencer is wrong about emotion, then he is wrong about expression, too.

Darwin then repeats Spencer's theory of the overflow of nerve force, embracing it:

> Mr. Spencer has also published a valuable essay on the Physiology of Laughter, in which he insists on "the general law that feeling passing a certain pitch, habitually vents itself in bodily action;" and that "an overflow of nerve-force undirected by any motive, will manifestly take first the most habitual routes; and if these do not suffice, will next overflow into the less habitual ones." This law I believe to be of the highest importance in throwing light on our subject. (1872, 9–10)

It is generally accepted that Darwin's third principle of expression, the principle of the direct action of the nervous system, is Darwin's attempt to articulate Spencer's principle of overflow in his own words (Fridlund 1994, 20). What has been overlooked, I believe, is that Spencer's principle presupposes his associationist theory of emotion, according to which emotions are really just trains of associations including the direct effects of overflow.

Darwin's praises of Spencer's theory, combined with his addition of the third principle of expression, show that Darwin continues to embrace the evolutionary associationist theory of emotion that he developed in 1838–1839 and that Spencer independently developed in 1855–1860. He praises Spencer's theory because it is also his theory. But for reasons I turn to now, Darwin is reluctant to claim credit for this theory in *Expression*. Spencer published it first. Since it is no longer his theory to present, Darwin says precious little about the nature of emotion in *Expression*, focusing instead on "following out the wonderfully complex results" of Spencer's theory (1872, 9).

In 1869, as his work on *Descent* and *Expression* was gearing up, Darwin published the fifth edition of *Origin*. This edition includes two important changes. First, it includes, for the first time, Spencer's gloss on natural selection, "the survival of the fittest" (Browne 2002, 312). Second, it replaces most instances of "my theory" with "the theory." Critics had charged Darwin with taking too much credit for natural selection. It turns out that others had come up with similar ideas before him, including Edward Blyth, Patrick Matthew, and William Wells, not to mention Alfred Russell Wallace (Browne 2002, 108–109). These criticisms stung, given Darwin's Victorian sense of honor, and he sheepishly changed "my theory" of natural selection to "the theory" of natural selection (Johnson 2012, 97).

I speculate that as Darwin was writing *Expression*, he was especially sensitive about taking credit for a theory that another had already published. Because Spencer published the theory before him, it could no longer be "my theory" of emotion. It could, at most, be "the theory" of emotion. Darwin was always possessive about his theories, however, so he couldn't help but write in a footnote beneath his summary of Spencer's theory, "my first MS. notes on the subject of expression bear the date of the year 1838" (1872, 10, f11). Spencer may have published first, but Darwin had the idea first.

3.8.2 Alexander Bain

In the years leading up to *Descent* (1871) and *Expression* (1872), Darwin studied another associationist philosopher, Alexander Bain.[32] Bain developed an associationist theory of mind in *The Senses and the Intellect* (1855, 1864) and then applied it to emotion in *The Emotions and the Will* (1859, 1865).[33] In a private correspondence, Darwin remarked that he found Bain's philosophy difficult to understand (Darwin 1873). It's possible that Darwin read John Stuart Mill's (glowing) review and Spencer's (critical) review of Bain's psychology, published in 1859 and 1860, respectively.[34] Yet when Darwin quotes Bain's works, he cites the heavily revised second editions, namely, *Senses* (1864) and *Emotions* (1865).[35]

Bain's theory of emotion is, in its official formulation, remarkably similar to Hartley's. Bain writes:

> The emotions, as compared with the sensations, are secondary, derived, or compound feelings. The emotion of wonder, for example, inspired by a human being of gigantic size, is the result of a comparison of the present

32 Swisher (1967, 40) discusses Bain's influence on *Descent* and *Expression*.

33 See Young (1970), Greenway (1973), Kennedy (2012), Rieber (2012), Harper (2019), and Dacey (2022) for scholarly treatments of Bain's philosophy.

34 One reason to think that Darwin read Mill's review is that a passage from *Expression* (1872, 367) resembles a passage from Mill's review (1978, XI:362). One reason to think that Darwin read Spencer's review is that he cites volume 1 of Spencer's *Essays: Scientific, Political and Speculative* (1861) in *Expression* (1872, 9n10), which includes Spencer's review of Bain.

35 Darwin's personal library did not contain the first edition of either. It contained the second editions of both and the third edition, published in 1875, of *Emotions*.

> impression on the sight with our recollected impressions of ordinary men and women.. . .
>
> [T]he theory of Emotion will consist in the statement of the general laws that regulate the concurrence, coalition, transfer, or aggregation of the primary feelings – the Muscular states and the Sensations; and the analysis of the several Emotions will be the account of the component feelings in each, and the mode of their composition under some of these general laws. Now, in addition to the adhesive property of Contiguous Association, two great principles operate in the concurrence of different feelings: the Law of Harmony and Conflict, and the Law of Relativity. (1865, 35–36)

In other words, emotions are trains of feelings ("sensations") and actions ("muscular states"), linked by general laws of association. Whereas Hartley's theory includes only contiguity, Bain mentions two other laws of association, "harmony and conflict" (which he also calls "similarity") and "relativity" (which he also calls "contrast"). Wonder, in Bain's analysis, occurs when a person sees a gigantic human being, contrasts them with normal-sized people previously encountered (via the principle of relativity), becomes mildly distressed by the unexpectedness of this sight (via the principle of conflict), and then becomes petrified, if only for an instant (1865, 45–46). Wonder denotes the whole unfolding episode, which may nonetheless be experienced in the moment as a simple feeling.

Although Bain's official formulation of his theory is clearly associationist, early critics complained that Bain fails to follow through on his promise to identify the associative links holding emotions together (Ribot 1874, 225–228). He gives detailed descriptions of many emotions, yet he often says nothing about how they are actually put together. We could say that Bain adopts an associationist theory of emotion in theory but not in practice.

Like Spencer, Bain attempts to identify the neuroanatomical basis of mental association (Rieber 2012). Hobbes claimed that once set in motion, the mind stays in motion. But what sets the mind in motion? Bain's "law of diffusion" states that feelings are accompanied by a diffusion of nerve energy throughout the brain and body.[36] He later clarifies that pleasure tends to be accompanied by an increase in muscle activity, while

[36] In his review of Darwin's *Expression*, Bain claims that Darwin's third principle is based on his own law of diffusion, yet Bain seems to have misunderstood Darwin's third principle. Darwin alludes to the law of diffusion in *Expression*, but objects that it "seems too general to throw much light on special expressions" (1872, 9). In his review of *Expression*, Bain claims that Darwin's third principle is a reformulation of the law of diffusion, which is "still more vague" (Bain 1873, 698). Darwin concedes that this reply is "perfectly just" (Darwin 1873).

pain tends to be accompanied by a decrease in muscle activity. We freeze in wonder because wonder can be a mildly painful experience (he claims).

Bain explores an interesting implication of his law of diffusion. He suggests that even if we attempt to suppress the outward expression of our feelings, the diffusion of nerve energy will find other, perhaps unexpected modes of expression. He writes:

> [I]t is possible, by force of will, to suppress the more prominent manifestations of feeling – namely, the movements depending on voluntary muscles. The organic effects, such as blushing, are beyond our power. The suppression of display may become habitual, but the feelings will still occur, although not unmodified by the refusal to allow them the natural vent. (Bain 1865, 8)

By "not unmodified," he means that the feelings will find alternative means of expression.

Darwin mentions Bain's law of diffusion in the introduction to *Expression* yet concludes that it "seems too general to throw much light on special expressions" (1872, 9). Bain suggests in his review of Darwin's *Expression* that his law of diffusion was the inspiration for Darwin's third principle, yet I think Spencer's law is the more probable influence. That's not to say that Bain had no influence on Darwin's *Expression.* I suggest that we can see this influence in Darwin's first principle, which he likely updated to incorporate Bain's ideas that the attempt to suppress an emotion will result in other, unexpected expressions.

> 1. *The principle of serviceable associated Habits.* – Certain complex actions are of direct or indirect service under certain states of the mind, in order to relieve or gratify certain sensations, desires, &c.; and whenever the same state of mind is induced, however feebly, there is a tendency through the force of habit and association for the same movements to be performed, though they may not then be of the least use. *Some actions ordinarily associated through habit with certain states of the mind may be partially repressed through the will, and in such cases the muscles which are least under the separate control of the will are the most liable still to act, causing movements which we recognise as expressive. In certain other cases the checking of one habitual movement requires other slight movements; and these are likewise expressive.* (1872, 28; emphasis added)

The non-italicized portion is purely Hartleian, and indeed Darwin had formulated these ideas much earlier in Notebook M (1838). However, the italicized portion is distinctively Bainian. Just as Bain mentions blushing as an expression that can't be suppressed, Darwin mentions "muscles which are least under the separate control of the will." And just as Bain claims that suppressed feelings will find other means of expression, Darwin supposes that the attempt to halt an expression will involve another expression.

In sum, although Darwin formulated his first two principles of expression in 1838, he revisited these principles in the late 1860s. Bain's work inspired him to add more to the first principle. Spencer's work inspired him to add a third principle.

3.9 Puzzles of *Expression*

My aim in this chapter has been to show that Darwin developed his theory of emotion and emotional expression amid a deep and prolonged engagement with associationist philosophers. In the late 1830s, Darwin wrestled with the ideas of David Hartley and Erasmus Darwin. Then, in the late 1860s, he took on Herbert Spencer and Alexander Bain. Hartley set the agenda. The elder Darwin, Spencer, and Bain pushed the Hartleian program forward. Darwin's *Expression* picks up where they leave off, attempting to use associationist principles to explain why we express our emotions as we do.

Darwin's engagement with associationist philosophy bookended his empirical studies of emotion and emotional expression. He formulated his first two principles of expression in the late 1830s. Then, in 1839–1840, he collected empirical data from his own child to support his theory. After a twenty-five-year hiatus, he began collecting more empirical data in the late 1860s, this time from observers scattered around the globe. But then his thoughts drifted back to the theoretical. After reading Spencer and Bain, he updated his first principle of expression and formulated a third. In both the beginning and the end of his inquiry, Darwin was steeped in associationist philosophy.

He sums up his theory of emotion with the words "yet emotions are results – are trains of thought long associated with action" (DAR LIB 404, 333). In other words, an emotion is the result of a thought or feeling becoming so strongly associated with an action that the one automatically triggers the other. In detailing the process by which thoughts and feelings become associated with actions – the express aim of *Expression* – Darwin spells out the psychological construction of emotion. Insofar as association is the core mechanism, Darwin deserves to be called an associationist theorist of emotion.

At this point we can solve the three puzzles of Darwin's *Expression* (from §1.5). The first puzzle is why Darwin appeals to heritable habits when his own theory of natural selection provides a superior alternative. The solution is that Darwin appeals to habits because he attempts to synthesize his theory of evolution with Hartley's associationist psychology. The latter posits habit as the basis of all thought, so Darwin's synthesis must likewise center on habit. To deny that emotions are (heritable) habits would be to reject, out of hand, the prevailing theories of psychology in Darwin's day.

To be clear, Darwin was wrong about heritable habits. We know now that habits are not heritable, at least not in the manner Darwin supposed, and so Darwin's analysis of emotions as heritable habits is incorrect – not incidentally or in some small respects, but deeply and fundamentally.[37] My point is just that we can't blame Darwin for getting this wrong, given that his main influences were wrong, too.

The second puzzle is why Darwin sidesteps emotion. Scholars claim that Darwin offers a theory of emotional expression, but not a theory of emotion. Furthermore, they argue that Darwin's theory of emotional expression, which regards many expressions as vestiges or spandrels, leaves no room for an evolutionary theory of emotion. Yet we have seen that these scholars are mistaken. Darwin does have a theory of emotion. It is not ruled out by his theory of expression; rather, it is tightly connected to this theory of expression.

Darwin theorizes that when a sensation becomes linked to an action, such that one triggers the other, the whole "train" or sequence becomes an emotion. He writes that "without abrupt flush, acceleration of pulse, or rigidity of muscles, – man cannot be said to be *angry*. He may have pain or pleasure these are sensations" (Darwin 2009, N 66). Strictly speaking, an emotion does not cause its expression. A sensation causes an expression, and when that happens, we call the whole episode an emotion. As Darwin puts it elsewhere, "ideas & bodily actions make the emotions" (Darwin 2009, N 89).

Darwin's analysis of emotion might be wrong, but it is not obviously wrong, and it very well could be right. His analysis resembles modern constructionist analyses, which deny that emotions are among the building blocks of the mind.[38] These analyses claim that emotions are constructed out of other ingredients via simpler psychological processes, including association. Constructionism is not without its critics, but it has its share of champions as well.

[37] However, recent studies suggest that some acquired experiences can influence the development of neural structures in offspring (e.g., Glover & Hill 2012; Dias & Ressler 2014; Dickson et al. 2018).
[38] I develop this claim in §4.6.

The third puzzle is why Darwin downplays communication. Recent theories posit that we smile when happy because smiling communicates happiness to observers, yet Darwin does not explain any particular expression in terms of its capacity to communicate emotion. The solution is that Darwin's associationism forecloses this possibility. If emotions do not exist independently of their expressions, then we cannot explain the expression in terms of the emotion, and certainly not in terms of an observer's concept of the emotion. We must explain why pleasant feelings lead to smiles without referencing joy at all. Darwin acknowledges that observers can infer emotions from expressions, but this is a result of the association, not its cause. Expressions do communicate emotions, but communication is not what these expressions are for.

Darwin's perspective on expression and communication may be wrong, but again, it is not obviously wrong. In contemporary debates, Darwin's view remains a live option.[39]

In this chapter we have explored Darwin's analysis of *emotion*. We have not yet explored his analysis of *emotions*. How many emotions do we have? What are they? How are they organized? That is the subject of the next chapter.

[39] I develop this claim in §5.7.

CHAPTER 4

Darwin's Emotions

4.1 From Emotion to Emotions

Charles Darwin developed a theory of emotion in 1838–1839, around the time that he discovered the logic of natural selection. He defines emotions as "trains of thought long associated with action" (DAR LIB 404, 333). You see a dog on the street. It reminds you of the dog that bit you. You remember how much it hurt. Wishing to avoid pain again, you turn and run away. This whole sequence of events – the sight, the memory, the pain, the desire, the action – constitutes the emotion of fear. Each event is linked to the next by an association, in some cases formed by a distant ancestor and then passed down to its many descendants.

Emotions are not, in Darwin's analysis, simple feelings. They are compound experiences of mind and body, constructed from diverse elements. He writes that "ideas & bodily actions make the emotions" (2009, N 89). In the sequence of events, a feeling causes an action. But the whole sequence is the emotion. To say that an emotion causes its expression is to commit a category mistake, to treat an emotion as a part rather than a whole. Emotions are not the causes of their expressions. Darwin is emphatic on this point: "emotions are results" (DAR LIB 404, 333).

A theory of *emotion* (singular) tells us what an emotion is and how emotions fit within a broader taxonomy of mind. In Darwin's theory, emotions are less like ideas and feelings (elements) and more like deliberations and daydreams (sequences). A theory of *emotions* (plural) tells us how many types of emotions there are and how they fit within a narrower taxonomy of emotions.[1] Are there eight primary emotions, which combine into twenty-four compounds? Are there five basic emotion families, each with dozens of members? Is contempt a subtype of anger, or a combination

[1] Odenbaugh (n.d.) helped me to appreciate the distinction between a theory of emotion (singular) and a theory of emotions (plural). Prinz (2012) articulates a basic framework for thinking about theories of emotions (plural).

of anger and disgust, or a separate emotion altogether? Is greed an emotion? A theory of emotions (plural) is a theory of emotion types, of anger, fear, sadness, joy, disgust, and so on.

Prior to these concrete questions of categorization, however, are two deeper philosophical questions. First, when we group emotions according to type – placing all the angers in one box, all the fears in another box, and all the joys in a third box – are these groupings natural or conventional (Bird & Tobin 2023, §1.1)? Is the difference between anger and fear more like the difference between hydrogen and oxygen or the difference between flowers and weeds? *Naturalists* say that there are at least some natural groupings of emotions. *Conventionalists* say there are no natural groupings of emotions, that all groupings reflect human interests and conventions.[2]

A naturalist theory of emotions (plural) should aspire to tell us exactly how many natural emotions there are and how they are arranged. This is because natural groupings are often thought to be categorical and hierarchical (Kuhn 2000; Ellis 2001). Think of biological classifications. Every organism is a member of a species and a genus. All members of an organism's species are also members of its genus. But not all members of its genus have to be members of its species. Natural emotions ought to be arranged similarly. Robert Plutchik (1962) proposes that there are eight primary emotions, organized into four opposing pairs, including fear, anger, joy, and sadness. These primary emotions combine into twenty-four compound emotions. Fear plus sadness equals despair. Anger plus joy equals pride. Paul Ekman (1992) proposes that there are at least six, and as many as twelve, categorically distinct emotion families, including fear, anger, joy, and sadness. The fear family includes trepidation, nervousness, dread, horror, and terror. The anger family includes annoyance, frustration, exasperation, bitterness, and fury. Philip Johnson-Laird and Keith Oatley (1989) categorize 590 emotion words into just five basic emotion families, including – you guessed it – fear, anger, joy, and sadness. A naturalist theory of emotions may not be able to provide a final count of natural emotion types, since research is ongoing and new emotion types may need to be added. But a final count should be the goal of any naturalist theory.

[2] The question of whether emotion groupings are natural or conventional should not be conflated with the question of whether the emotions grouped together are homogeneous or heterogeneous. Barrett (2006) cites the heterogeneity of angers as evidence that "anger" does not denote a natural grouping of emotion, but Griffiths (1997) argues that natural groupings of emotions can be heterogeneous, as long as there is a mechanism that limits variation. Owing to their arbitrariness, conventional groupings can be homogeneous or heterogeneous.

A conventionalist theory of emotions, by contrast, may deny that there is any definitive way of counting or taxonomizing emotion types.[3] This is because conventional groupings, unlike natural groupings, can be vague and disorderly.[4] The difference between anger and fear may be like the difference between art and entertainment; there are clear exemplars on each side but many instances that occupy the murky space between. Not only that, but what counts as anger or fear can differ from person to person, from culture to culture, and from era to era.[5] William James puts the point like this:

> If one should seek to name each particular one of [the emotions] of which the human heart is the seat, it is plain that the limit to their number would live in the introspective vocabulary of the seeker, each race of men having found names for some shade of feeling which other races have left undiscriminated. If we should seek to break the emotions, thus enumerated, into groups, according to their affinities, it is again plain that all sorts of groupings would be possible, according as we chose this character or that as a basis, and that all groupings would be equally real and true. (James 1890, 485; quoted in Johnson-Laird & Oatley 1989, 82)

Conventionalist theories of emotion tend to be less interested in counting and taxonomizing emotion types and more interested in deconstructing seemingly "natural" groupings of emotions. Some historians argue that what we mean by "anger" today is very different from what the Victorians meant by "anger" in the nineteenth century, and that these distinct groupings reflect social, economic, and political conditions (Stearns & Stearns 1986). Some anthropologists argue that the Western concept of "anger" is foreign to some other cultures (Briggs 1970; Lutz 1988). In each case, the claim is that the word "anger" picks out different conventional groupings of emotions, depending on who utters it and when.

The first philosophical question is whether groupings of emotions are natural or conventional. The second is whether, to group emotions, we

[3] Of course, a conventionalist theory can affirm that there is a definitive way of counting and taxonomizing emotion types, as long as the count and arrangement are determined by convention and not by nature. I will examine conventionalist theories that deny this, however.

[4] See Russell (1991), Russell & Fehr (1994), Russell (2003), Fehr & Sprecher (2009), and Barrett (2017b) for arguments that emotion concepts denote individually and culturally variable prototypes. They claim that we call an emotion "anger" just in case it more closely resembles our prototype of anger than our prototypes of other emotions. These emotions do not, however, form natural groupings that are categorical or hierarchical.

[5] Among the best treatments of cultural and historical differences in emotion concepts are Stearns & Stearns (1985), Lutz (1988), Stearns (1993; 1994), Reddy (2001), Dixon (2003), Rosenwein (2006; 2015; 2020), Frevert (2011), Illouz (2013), and Plamper (2017). See also Wierzbicka (1992), Reisenzein (1995), and Ng et al. (2019).

must assert the existence of the boxes we place them in (Bird & Tobin 2023, §1.2). *Realists* claim that in addition to all the individual episodes of anger, there is something else that exists, namely, the general kind, anger. This kind may have an "essence" – a set or cluster of properties that marks any given episode of emotion as one of anger, specifically.[6] *Nominalists* deny that emotion kinds exist over and above all the individual episodes of emotion. There is no anger, but there are of course many angers. These angers may resemble one another in various ways, leading us to group them together, but they do not share an essence.[7]

Nineteenth-century philosophers tended to run these two philosophical questions together.[8] Thomas Brown developed a theory of emotions that combines conventionalism and nominalism. "In nature," he writes, "there are no classes. We give the names of fear, and tender pleasure, to various feelings; but there exist only particular feelings of fear, of pleasure" (Brown 1798, 321). He reasons that if there are no natural groupings of emotion (conventionalism), then only individual emotions exist (nominalism). In his *Lectures on the Philosophy of the Human Mind*, he names dozens of different emotions – ranging from melancholy and hate to humility and remorse – but insists that his groupings are arbitrary assemblies of motley emotions (1824, II:52, p. 255). An episode of joy may have more in common with an episode of anger than a second episode of joy, which may have more in common with an episode of fear. We could just as well classify these emotions as two angers and two fears, or give each its own name.

Herbert Spencer developed an opposing theory of emotions that combines naturalism with realism. Building on his (pre-Darwinian) theory of evolution, he argues that there are at least four natural emotions, grouped according to biological function. Anger evolved as an adaptive response to competition, while fear evolved as an adaptive response to danger (Spencer 1855, §200). Aesthetic appreciation is an adaptive response to novelty (Spencer 1855, §201), while love is an adaptive response to mates,

[6] The classical view of essences is that they are sets of intrinsic properties shared by all members of a kind. A recent, alternative view of essences is that they can be homeostatic property clusters (Griffiths 1997; Boyd 1999). Even if there is no property that all angers share, anger can still have an essence in the form of a mechanism that limits variation among the properties of individual angers.

[7] Since realists can equate emotion kinds with homeostatic property clusters, nominalists may have to deny that there are emotion-specific mechanisms limiting variation among the properties of emotion types. See Craver (2009), Slater (2015), Tobin (2017), and Chakravartty (2023) for relevant discussions.

[8] Contemporary metaphysicians recognize that the naturalist/conventionalist debate swings freely of the realist/nominalist debate. See Khalidi (2016) and Bird & Tobin (2023, §1.2).

offspring, and possessions (Spencer 1855, §202). Spencer hypothesizes that each emotion is localized in a different part of the brain, where dedicated nerves coordinate sensation and response (Spencer 1855, §204). These emotions form natural groupings (naturalism) because they share essential structures and functions (realism). We speak of other emotions like melancholy and gratitude, but these latter emotions may not be as natural or as primitive as anger and fear.

To sum up their disagreement: Brown claims that the concepts of "anger" and "fear" impose human structure on natural flux, while Spencer claims "anger" and "fear" carve nature at its joints. Brown suggests that we can arrange emotions however we want, since all arrangements are ultimately arbitrary. Spencer suggests that the goal of emotion science is to discover nature's hidden order.

What makes the Brown–Spencer debate especially interesting is that despite their opposing theories of *emotions* (plural), they embraced remarkably similar theories of *emotion* (singular). Both were associationists who analyzed emotions as trains of thought and action. You see a dog. You recall a painful bite. You run away. The whole sequence is the emotion. Brown and Spencer agree on this much, although they disagree on what follows. Spencer claims that this emotion naturally belongs in the same box as other fears, based on a shared essence. Brown claims that there are no boxes, or that any grouping of this emotion with others is merely a matter of convention or convenience. The Brown–Spencer debate was an internal debate among associationists.

Charles Darwin was no stranger to this debate. He studied Brown's theory of emotions. He studied Spencer's theory of emotions. In *Expression* (1872), he weighs in on their debate. If you had to guess Darwin's position before reading *Expression*, then you might guess that Darwin sides with Spencer. After all, Darwin also develops an evolutionary theory of emotion, tracing our emotional expressions back to the lives of our simian ancestors. It's a good guess, but I shall argue that it's wrong. Darwin in fact sides with Brown. He embraces conventional groupings of emotion (conventionalism), which are vague and disorderly, and he rejects the postulation of emotion kinds over and above individual episodes of emotion (nominalism). Darwin was a nominalist about species – *Homo sapiens* doesn't exist; only individual humans exist – and he was a nominalist about emotions, too – *anger* doesn't exist; only individual angers exist. Darwin does not offer a clear count or taxonomy of emotions, since he does not believe that emotions are the sorts of things that can be rigorously classified. What matters to Darwin is the simple psychological process, association, which produces all the wonderful variation we find in our rich emotional lives.

In this chapter I will situate Darwin's *Expression* within two debates: first, the nineteenth-century debate between Brown and Spencer over emotion types; second, its modern analogue, the twenty-first-century debate between basic emotions theory (BET) and psychological constructionism (PC). BET affirms that there are natural groupings of emotions, corresponding to natural kinds. PC denies that emotions are natural kinds. I will argue that although Darwin is often heralded as the forefather of BET, his theory of emotions paved the way for PC.

4.2 Brown's Emotions

Thomas Brown was a Scottish philosopher who lived from 1778 to 1820.[9] He published three important works of philosophy. The first, a critique of Erasmus Darwin, was published in 1798.[10] The second, a defense of David Hume, was published in 1805. And the third, his lectures on the philosophy of mind, were published in 1820, the year of his untimely death.

The historian Thomas Dixon has dubbed Brown "the inventor of the emotions" (Dixon 2003, 109). The claim sounds preposterous at first. But Dixon's point is that Brown was the first theorist to consistently use the word "emotion" in its modern sense, denoting a genus of feelings that includes fear, anger, sadness, and the like. Because the term "passion" had religious connotations and overtones, Brown proposed "emotion" as a secular, scientific alternative (Dixon 2003, 133–134). If a science of feeling is possible, then it will be a science of the emotions rather than a science of the passions.[11]

Although Brown wallows in obscurity today, he was an intellectual giant in the first half of the nineteenth century. His *Lectures on the Philosophy of the Human Mind* (1824) went through twenty editions in twice as many years (Dixon 2003, 111). The philosopher J. D. Morell expressed a common sentiment when writing of the excitement that Brown's writings inspired in him:

9 For scholarly treatments of Brown's associationism, see Mackintosh (1837), Warren (1921), Landes (1926), Dixon (2003), and Dacey (2015).

10 See King-Hele (1999, 314–315) for a historical account of Erasmus Darwin's reaction to this critique.

11 Brown's idea of a science of the mind was, however, purely introspective. He saw little value in empirical studies of the brain. As Dacey points out, Brown "refuses to speculate on physical or physiological correlates to psychological phenomena" (Dacey 2015, 34).

> I next betook myself to the Lectures of Dr Thomas Brown, hoping to find there to satisfaction I required. In this hope I was not for the time disappointed. The style was so captivating, the views so comprehensive, the arguments so acute, the whole thing so complete, that I was almost insensibly borne along upon the stream of his reasoning and eloquence. Naturally enough I became a zealous disciple; I accepted his mental analysis as almost perfect; I defended his doctrine of causation; with him I stood in astonishment at the alleged obtuseness of Reid. (Morell 1846, I:vi–vii; quoted in Dixon 2003, 112; italics removed).

Dixon argues that this whimsical praise translated into real scientific influence: Brown's work shaped emotions research in the English-speaking world for nearly a century (Dixon 2003, 135–179).

Brown's philosophy of mind can best be summed up as a Humean twist on associationism.[12] Associationists posit association as a fundamental psychological mechanism. Hume, of course, was skeptical of causation beyond constant conjunction. Putting these two ideas together, Brown (1805) argues that we cannot know the underlying causes of our thoughts, only that one thought has consistently followed another. As Margaret Landes puts it, "The 'gentle force' by which one idea attracts other ideas is not observed fact but hypothesis, introduced by the associationists to explain the fact that sensations and 'ideas' are experienced as related" (1926, 450). Or as Brown himself puts it, "if we trust the evidence of our consciousness, which affords the only evidence, we have no knowledge of any intermediate process that can have the name association" (1820, 2:334; quoted in Landes 1926, 452).

In Brown's view, a science of the mind should simply describe patterns of thought without attributing unobserved causes. In an effort to clarify his position, Brown uses the word "association" to denote a hypothetical mechanism of thought, of which he is skeptical, and the word "suggestion" to denote the mere fact that one thought followed on the heels of another. Brown thus develops a theory of "suggestion" rather than a theory of "association" per se. Granted, the word "suggestion" smacks of an unobserved force even more so than "association" (Mackintosh 1837, 347). But the difference between unobserved causation and constant conjunction is plain enough (Landes 1926, 453).

12 See §2.4 for my account of Hume's contributions to associationism and why I classify him as a quasi-associationist about emotion.

Despite his criticisms of associationism, many scholars classify Brown as an associationist.[13] After all, he denies that there are innate faculties of the mind. He affirms that all ideas derive from experience. And he attempts to reduce all complex mental phenomena, including emotions, to trains of simpler ideas. It's perhaps safest to say that Brown is an off-brand associationist.

I shall focus on Brown's conventionalist, nominalist theory of emotions. We find an early formulation of this theory in his critique of Erasmus Darwin (1798). Recall that Erasmus Darwin's *Zoonomia* (1796) presents an associationist theory of emotion and emotional expression.[14] Emotions are trains of associations. You see a dog, you remember that time a dog bit you, you remember how much it hurt, you wish to avoid this pain, so you turn and run away. This whole sequence of events, each joined to the next by the power of association, constitutes the emotion of fear. Anticipating his grandson's analysis of expression, Erasmus argues that many emotional expressions are movements that happened to accompany strong sensations in the past. Given their association, any time we have those strong sensations again, we may automatically produce the same movements again, even though they may serve no purpose.

Brown agrees, to an extent, with this analysis of emotion. He affirms that emotions are trains of associations.[15] However, he denies that we can understand the causal mechanism underlying these trains. Instead, he claims that we can only know the sequence of thoughts, feelings, and actions that occur in them. Yet – and this is why many view him as an

[13] Mackintosh (1837), Warren (1921), Titchener (1928), and Dacey (2020) classify Brown as an associationist. Landes (1926) argues that Brown is not an associationist.

[14] I presented Erasmus Darwin's theory in §3.2.

[15] Brown writes that emotions "may be defined to be vivid feelings, arising immediately from the consideration of objects, perceived, or remembered, or imagined, or from other prior emotions" (I.16, p. 373). Thomas Dixon remarks that this definition is exceedingly vague and thus that "Brown failed to provide a clear definition of 'emotions'" (Dixon 2003, 125). I disagree with Dixon's assessment. I view Brown's definition as a clear statement of agreement with the associationist tradition. Locke similarly analyzed emotions as feelings of pleasure or unease upon the thought of an object. Fear, for instance, is an "uneasiness of the mind, upon the thought of future evil likely to befall us" (1894, II.20.10, p. 305). Neither Brown nor Locke identifies emotions with simple feelings, as one reading of these definitions would suggest. Rather, both identify emotions with *sequences* or *trains* of thoughts and feelings (see §2.3 for my defense of this interpretation). Fear is the thought of future evil followed by the feeling of unease (which itself is a faint copy of the unease previously felt in the presence of evil). Since the experience of evil leads to a feeling of unease, resulting in a mental association, the subsequent thought of impending evil sparks its associate. As Brown's definition makes clear, this sequencing can be iterative. A thought can lead to a feeling, which can lead to another thought and feeling, and then others. Neither Brown nor Locke emphasizes that the train of associations leads also to expressive behaviors, but neither would deny this either.

associationist – he accepts that we can describe the relations between these events using principles of association – er, suggestion.

Brown disagrees emphatically with Erasmus's view that emotion kinds exist in nature: "It may be objected to the whole of Dr. Darwin's theory of natural signs, that it is founded on artificial language, and not on nature" (1798, 321).[16] Brown insists that, "In nature, there are no classes. We give the names of *fear*, and *tender pleasure*, to various feelings; but there exist only particular feelings of fear, of pleasure" (1798, 321). In other words, Brown objects to Erasmus Darwin's assumption that emotions are naturally grouped into categories like "fear" and "tender pleasure." These groupings are not natural at all. They are conventional. Brown then pivots from conventionalism to nominalism, suggesting that we ought to speak of "fears" instead of "fear," since the plural form helps us to focus on the individual episodes as opposed to the kind, since there is no real kind.

Brown develops his conventionalism further in his *Lectures* when he first embarks on the topic of emotion:

> In now entering on the consideration of that order of our feelings, which I have comprehended under the name of Emotions, it may seem doubtful whether it would be more expedient to treat of them simply as elementary feelings, or in those complex forms in which they usually exist, and have received certain definite characteristic names that are familiar to you. This latter mode appears to me, on the whole, more advisable, as affording many advantages, direct and indirect, and allowing equally the necessary analysis in each particular case. If I were to treat them only as elementary feelings, they might be classed under a very few heads, – the whole, as I conceive, or certainly, at least, the greater number of them, under the following: Joy, grief, desire, astonishment, respect, contempt, and the two opposite species of vivid feelings, which distinguish to us the actions that are denominated vicious or virtuous. But, though the vivid feelings, to which we give these name, may, from their general analogy, admit of being comprehended in this brief arrangement, *it must be remembered, that brief as the vocabulary is, it comprehends feelings, which, though analogous, are still not precisely the same, – that the single word joy, for example, expresses many varieties of delightful feelings, the single word desire, many feelings, which, in combination with their particular objects, are so modified by these, as to appear to us, in their complex forms, almost as different as any other feelings of our mind which we class under different names.* It is in their complex state that they impress themselves most strongly on our observation in others, and form, in ourselves, all that renders most interesting to us the present and the future,

[16] Brown attributes this view to Erasmus. I am not convinced, however, that Erasmus actually held this view.

and all that is most vivid in our remembrances of the past. (Brown 1824, II.52, p. 255; emphasis added)

In this lengthy passage, Brown repeats his skepticism of natural groupings of emotion. He writes that we could try to identify a core set of "elementary feelings," including "Joy, grief, desire, astonishment, respect, [and] contempt." The problem with this approach is that each of these names refers to a wide range of individual emotions that differ from each other as much as they differ from individual emotions with different names. In Brown's words: "the single word joy … expresses many varieties of delightful feelings … which, in combination with their particular objects … appear to us … almost as different as any other feelings … which we class under different names."

Today this argument against naturalism is called the variability problem.[17] If there are emotion kinds, then there should be physiological or phenomenological features – "fingerprints" – that all members of the kind share. However, there are no such features. For any supposedly essential feature of, say, sadness, there are episodes of sadness that lack that feature, and there are episodes of other emotions that exhibit that feature. Therefore, emotions do not form natural groupings.

Given his skepticism of any natural, objective groupings of emotions, Brown advocates an alternative approach: Discuss emotion types in their order of familiarity. We are most familiar with emotions like "cheerfulness" and "melancholy," so the discussion ought to begin there. He discusses altogether some thirty or forty emotions, organized as shown in Table 4.1.

Brown classifies emotions – arbitrarily, he insists – into three families, each divided into two genera. The first family of emotions are "immediate" in the sense that they are directed toward present states of affairs. Sympathy, for instance, is directed toward someone else's occurrent emotions. Immediate emotions are divided into those that do not involve moral judgment and those that do. Brown claims that cheerfulness, melancholy, and wonder are amoral emotions, whereas sympathy, pride, and humility are moral emotions.

The second family of emotions are "retrospective" in the sense that they are directed toward past states of affairs. Brown claims that anger and

[17] Barrett (2006) presents the variability of emotions as evidence that emotion kinds are not natural kinds. However, Scarantino (2015), Loaiza (2021), and Majeed (2022) challenge the objection. They argue that the variability of emotions is consistent with the claim that emotion kinds are natural kinds. Brown may be wrong that the variability of emotions supports his nominalism, yet my point is simply that he *thinks* that the variability of emotions supports his nominalism.

Table 4.1 *Brown's emotions*

Immediate		Retrospective		Prospective	
Amoral	Moral	Other-Directed	Self-Directed	Desire for	Fear of
Cheerfulness	Love	Anger	Regret	Existence	[Death]
Melancholy	Hate	Gratitude	Gladness	Pleasure	[Pain]
Wonder	Sympathy		Remorse	Action	[Inaction]
Languor	Pride		[Pride]	Society	[Loneliness]
Beauty	Humility			Knowledge	[Ignorance]
[Ugliness]*				Ambition	[Powerlessness]
Sublime				Avarice	[Poverty]
Ludicrous					

* Emotions in brackets are not named as such but referred to as the "opposite" of another entry on the list. For instance, when he discusses "beauty and its opposite," I have labeled the latter as "ugliness." And when Brown discusses the desire for existence and the fear of its opposite, I have labeled the latter the fear of "death."

gratitude are reactions to bad or good things that have already happened to you. Anger and gratitude comprise the first genus of retrospective emotions, namely, those that are directed toward the actions of another. The second genus of retrospective emotions includes those directed toward one's own actions. Regret, for instance, concerns one's past actions.

The third family of emotions are – you guessed it – "prospective" in the sense that they are directed toward future states of affairs. Brown divides this family into two genera: desire and fear. Desires include the desire to continue living, the desire for pleasure, the desire for company, and even the desire for knowledge. Fear includes the fear of death, the fear of pain, the fear of being alone, and the fear of ignorance.

So far, I've presented Brown's arguments for conventionalism followed by his list of conventional emotion types, organized by order of familiarity. I will later argue that Darwin follows Brown's lead. In *Expression*, Darwin states that emotion types are conventional and he discusses more than fifty emotion types, organized according to Darwin's convenience. Before turning to Darwin, however, I will present Spencer's naturalist, realist alternative to Brown's theory.

4.3 Spencer's Emotions

Herbert Spencer's *Principles of Psychology* (1855) has been called the first great work of evolutionary psychology (Leslie 2006). It bears repeating that although Charles Darwin discovered natural selection, the mechanism of evolution, a number of philosophers and scientists already believed that organisms adapt and evolve, including Charles's grandfather, Erasmus Darwin, Jean-Baptiste Lamarck, and Herbert Spencer. Erasmus and Lamarck speculated that more complex mental capacities evolved from less complex ones, but Spencer was really the first to offer rich evolutionary explanations for particular psychological phenomena. In what follows I will present Spencer's method of reconciling an associationist theory of mind with a naturalist, realist theory of emotions (plural).

Spencer's *Principles of Psychology* advances a "synthetic" philosophy of mind. It attempts to reconcile two theoretical approaches to the mind that were viewed as incompatible. The first is associationism, which (prior to Spencer) held that the mind begins as a blank slate, and that all mentality consists of sensations and associations among sensations. The second approach to the mind is phrenology. Today people think of phrenology as the debunked pseudo-science of correlating bumps on the skull with criminality, but Spencer thought of it more charitably as the attempt to identify neural correlates of mental processes (Young 1970). While associationism appealed to high-minded philosophers, phrenology appealed to empirically minded physiologists. The two approaches were thought to be incompatible because the phrenologists posited innate mental faculties, corresponding to different structures in the brain, while the associationists rejected innate faculties, insisting that the various mental faculties are acquired in the course of life experience.

Spencer's synthesis of these two approaches is rather ingenious. He draws on two of Lamarck's laws. The first is the law of use and disuse. Habitual use of an organ enlarges it; habitual disuse shrinks it. The second is the law of inheritance. Parents with large organs give birth to offspring with large organs; parents with small organs give birth to offspring with small organs. Lamarck famously applied these two laws to explain the giraffe's long neck. The more a giraffe stretched its neck to reach high branches, the longer its neck became. The giraffe then passed this trait down to its offspring.

Spencer extended these ideas to the mind. Applying the law of use and disuse, he hypothesizes that repeated associations of ideas strengthen neural connections. In other words, thinking two thoughts in sequence over and over again will enlarge a specific part of the brain. If, every time you hear the word "salt," you think "pepper," then a particular neural connection is going to get stronger and stronger. Ideas can also become decoupled over time, leading to a weakening of the neural connection.

This idea goes beyond simple supervenience – that there is no mental change without a physical change – by positing a law-like relationship between mental and physical changes.

Next, applying the law of inheritance, Spencer hypothesizes that acquired neural structures are heritable. In other words, if you learn to associate "salt" and "pepper," strengthening that neural connection, then your offspring may inherit that association. Upon learning what salt and pepper are, they may be inclined to think that they go together, even without having learned this connection through experience. In this way, Spencer suggested that one generation's a posteriori knowledge can become the next generation's a priori knowledge.

Spencer's synthesis of associationism and phrenology is clever, albeit incorrect by modern standards.[18] In his view, the associationists were right to think that all mental structures are acquired, but wrong to think that all mental structures are acquired *by each individual.* The phrenologists were right to think that there are innate mental capacities, but wrong to think that these capacities are immutable. *The* mind starts as a blank slate, but *your* mind does not. Your mind has been preprogrammed with some of your ancestors' experiences.

Spencer applies this analysis to emotion. Emotions are first formed through experience. An animal learns that fighting leads to pain, and so anticipates pain from fighting. The animal's offspring then inherits the association between pain and fighting. It anticipates pain from fighting, even before it gets into its first fight. That's the origin of fear.

> To have in a slight degree those psychical states accompanying the reception of wounds, those which express themselves in cries, those which are experienced during flight, is to be in a state of what we call fear. (1855, §200)

To experience fear is to simulate, albeit less vividly, what your ancestors experienced in a bloody fight. They thought danger; you think danger. They felt pain; you feel pain. They screamed; you scream. The elements go together now because they happened together then. Events that occur together in time become linked in the mind. Fear is the product of association.

Spencer refers to fears as *protective* passions. Angers, in turn, are *destructive* passions:

[18] The Modern Synthesis marked the death of Spencer's account of evolution by inheritance of acquired characteristics (Huxley 1942; Mayr 1942; Stebbins 1966; Bowler 2003). I discuss the Modern Synthesis further in §6.2.

> The destructive passions are shown in a general tension of the muscular system, in gnashing of the teeth and protrusion of the claws, in dilated eyes and nostrils, in growls; and these are weaker forms of the actions that accompany the killing of prey. To such objective evidences, everyone can add subjective ones from his own experience. Everyone can testify that the psychical state which . . . we call anger, consists of mental representations of the actions and impressions which would occur while inflicting some kind of pain on another. (1855, §200)[19]

Again, anger is nothing more than the conjuring of ancestral memories in the present. The memories include what our ancestors felt and how they behaved in response.

Spencer then turns to the *passive* passions of aesthetic appreciation. In contrast to fear and anger, Spencer claims that aesthetic appreciation develops in complexity over the course of a lifetime. An adult's experience of aesthetic appreciation is much richer and more sophisticated than a child's:

> The infant taken into the midst of mountains, is totally unaffected by them; but is delighted with the small group of attributes and relations presented in a toy. The child can appreciate, and be pleased with, the more complicated relations of household objects and localities, the garden, the field, and the street. But it is only in youth and mature age, when individual things and small assemblages of them have become familiar and automatically cognizable, that those immense assemblages which landscapes present can be adequately grasped, and the highly aggregated states of consciousness produced by them, experienced. (1855, §201)

The reason is that aesthetic appreciation hinges upon the contrast between familiarity and unfamiliarity. We enjoy unexpected ideas in familiar situations and familiar ideas in unexpected situations. Too much familiarity, and we become bored. Too little familiarity, and we become overwhelmed. In either case, we disengage. But when the balance of familiarity and unfamiliarity is just right, our attention holds. We become fixated. Our bodies become motionless, but our minds bustle with trains of associations.[20]

The final emotions that Spencer analyzes are the *erotic* passions of love. Spencer argues that there are many types of love, but that they are all connected by association. There is the affection that occurs between

[19] Spencer conflates aggression and anger, predation, and competition. Lorenz (2002) distinguishes these more clearly.

[20] See Townsend (1993) for an analysis of Joseph Priestley's Hartleian theory of aesthetic appreciation, which closely resembles Spencer's. See also Jauss (2006) for important context.

friends. Then there is the "admiration, respect, or reverence" that we may feel toward a stranger (1855, §202). Following this is the love of approbation, or the desire to be approved of by friends and strangers alike. Spencer adds the loves of possession and freedom. Intriguingly, Spencer argues that romantic, sexual love is the most complex and powerful emotion of all, since it combines all these other forms of love:

> the passion which unites the sexes ... is habitually, but very erroneously, spoken of as though it were a simple feeling; whereas it is in fact the most compound, and therefore the most powerful, of all the feelings. Added to the purely physical elements of it, are first to be noticed those highly complex impressions produced by personal beauty; around which are aggregated a variety of pleasurable ideas, not in themselves amatory, but which have an organized relation to the amatory feeling. (1855, §202)

Spencer focuses on just four emotion types: fear, anger, aesthetic appreciation, and love. These are not *conventional* groupings of emotion. They are *natural* groupings of emotion. They purport to carve nature at its joints. Spencer identifies each of these emotion types with a particular biological function. Fear is related to defense, or the need to protect oneself from predators and rivals. Anger is related to offense, or the need to attack rivals and prey. Aesthetic appreciation is related, ultimately, to exploration. We seek the unfamiliar while taking comfort in the familiar. Finally, love is related to reproduction, not merely in the narrow sense of sexual intercourse and conception, but in the much broader sense of having the freedom, resources, and relationships necessary for raising children into successful adults. We experience these four emotions today, not because Victorian culture says we do, but because evolution has molded and preserved them. (A conventionalist might argue that Spencer focused on these four emotions *precisely because* of their centrality to Victorian culture. My aim here is not to evaluate whether Spencer's view is right or wrong, but simply to present his view.)

Spencer further hypothesizes that his four natural emotions are enacted by dedicated neural structures, connecting perceptions of major life circumstances with distinctive behavioral patterns.

> As every student of the nervous system knows, the combination of any set of impressions, or motions, or both, implies a ganglion in which the various nerve-fibres concerned are put in connection.... [I]t follows that every emotion implies some portion of nervous structure by which its various elements are united. (1855, §204)

He calls these dedicated structures "the seat of the emotion" (1855, §204). Recall that Spencer wishes to synthesize associationism with phrenology.

From associationism we can learn how emotions aggregate from elemental states. From phrenology we can learn where each emotion is located in the brain.[21]

Finally, Spencer implies that various expressions are universal and that they can also be found in other animal species. He writes that anger is generally expressed in "a general tension of the muscular system, in gnashing of the teeth and protrusion of the claws, in dilated eyes and nostrils" (1855, §200). Humans cannot protrude their claws, of course, but they can become tense, gnash their teeth, and flare their nostrils – and these are indeed among the hypothesized universal expressions of anger. Spencer speculates that emotions emerged "in the course of the general evolution of Life" when a species has achieved "a considerable degree of complexity," such that "conflicting tendencies among some of the psychical changes arise" (1855, §199). He adds that this was the same point at which memory and reason emerged (1855, §199). Spencer does not get any more specific, but presumably all mammals have reached this stage, and perhaps most vertebrates and some invertebrates.

In sum, whereas Brown defends a conventionalist, nominalist version of associationism, Spencer defends a naturalist, realist version of associationism. Assuming that associations can be inherited, and assuming further that some of our emotions have been passed down over many thousands of generations, some of our emotions belong to emotion kinds that have been inscribed into the folds of the mammalian brain.

4.4 Darwin's Emotions

Charles Darwin agrees with many aspects of Spencer's evolutionary associationism, yet I shall argue that Darwin rejects Spencer's naturalist, realist theory of emotions. Instead, Darwin embraces Brown's conventionalist, nominalist theory.

I begin by reviewing Darwin's knowledge of Spencer and Brown. We know that Darwin studied Spencer's theory of emotions, as I documented previously (§3.8.1). We also know that Darwin read summaries of Brown's theory of emotions. John Abercrombie's *Inquiries Concerning the Intellectual Powers and the Investigation of Truth* (1838) includes a summary of Brown's theory of association and emotion. Darwin marked up this chapter in his personal copy of the book, and then scribbled on the last two blank pages, "Nothing for Specis [*sic*] book"

[21] See Young (1970) for a detailed analysis.

but also "Origin of shame & blushing, fear & anger mixed?"[22] Similarly, Sir James Mackintosh's *Dissertation on the Progress of Ethical Philosophy* (1836) ends with a summary and evaluation of Brown's philosophy.[23] The summary references Brown's critique of Erasmus Darwin, where he first developed his conventionalist, nominalist theory of emotions.[24] Darwin marked up these pages in his personal copy of Mackintosh's *Dissertation.*

After reading these summaries of Brown, Darwin added Brown to a list of philosophers he either read or intended to read more closely:

> "Adam Smith Moral Sentiments" much on life & character
> "'Humes Dissertation on the Passions."
> "Hartley" I should think well worth studying –
> "Thomas Brown" on Association worthy of close study. – full of practical observations.
>
> – (Darwin 2009, N, 184)

Darwin's library may have contained an 1828 edition of Brown's *Lectures on the Philosophy of the Human Mind*, although this copy appears to have been lost, and so it is not possible to see if Darwin marked it up.[25]

Having established that Darwin was familiar with Brown's and Spencer's theories of emotions, I will now argue that Darwin takes Brown's side in the debate.

The first reason to think that Darwin is a nominalist about emotions is that he is a nominalist about species.[26] In the eighteenth century, philosophers of science debated whether biologists' classifications carve nature at its joints or impose human conventions onto nature. Georges-Louis Leclerc, Comte de Buffon, argued that Carl Linnaeus's famous binomial

[22] Manier (1978, 59–60) and Dixon (2003, 161) discuss Darwin's annotations in his personal copy of Abercrombie's *Inquiries.*

[23] The preface of the *Dissertation*, written by William Whewell, observes that Mackintosh's own views are thoroughly Brownian. Darwin was personally acquainted with both Mackintosh and Whewell and referred to them in his autobiography as the best conversationalists on deep topics he ever met (Darwin 1958, 66).

[24] This citation must have sparked Charles Darwin's curiosity, especially since Darwin had reread his grandfather's book the previous year. However, he did not make a visible mark on this page, and I have not found evidence that Darwin ever obtained or read this work. Dixon (2003, 161) discusses Darwin's engagement with Mackintosh's *Dissertation.*

[25] I cannot prove that Darwin followed through on his plan to study Brown more closely. However, there are passages from Darwin's *Expression* that resemble Brown's *Lectures*, which lead me to believe that Darwin did. Compare Brown (1824, LIII: 268–269) with Darwin (1872, 278, 289) and Brown (1824, LXIII: 428–429) with Darwin (1872, 239). In each pairing, the author draws similar distinctions among varieties of emotions in order to emphasize the variety within purported emotion types.

[26] Lennox (2019, §2.3.4) presents this as the most plausible interpretation. But see Ghiselin (1969), Beatty (1985), and Stamos (2007) for alternative interpretations.

classifications were unnatural. "The basic concepts employed by taxonomists, such as family and genus, were for [Buffon] mere human inventions" (Jordanova 1984, 13).[27] Buffon believed that all of nature exists along a continuum. In his *Natural History* (2007/1749–1767) he discusses organisms in order of familiarity. He begins with those that are closely connected with humans and then branches outward to the more exotic (Jordanova 1984, 13).

Near the beginning of *Origin*, Darwin writes: "I look at the term species, as one arbitrarily given for the sake of convenience to a set of individuals closely resembling each other" (1859, 52). Individual dogs exist, including Rex, Fido, and Gunner. These individuals resemble each other in various respects, so we lump them into the species *Canis familiaris*. But we do this "arbitrarily" and "for the sake of convenience." We shouldn't think that the species is a real thing, which exists over and above the members of the species. Darwin repeats this point again near the end of *Origin*: "In short, we shall have to treat species in the same manner as those naturalists treat genera, who admit that genera are merely artificial combinations made for convenience" (1859, 485). Some naturalists regarded species as real but genera as conventional. Darwin seems to be saying that species, like genera, are conventional categorizations.

Ernst Mayr (1994) reads Darwin as a nominalist.[28] He claims that one of Darwin's greatest contributions to biology is that he replaced "typological thinking" with "population thinking." If we think of species as *types*, then we assume that all members share an essence and that this essence is real (Mayr 1994, 158). But if we think of groups as *populations*, then we recognize that "Individuals . . . form populations of which we can determine only the arithmetic mean and the statistics of variation. Averages are merely statistical abstractions; only the individuals of which the populations are composed have reality" (Mayr 1994, 158). There is some debate as to whether population thinking entails nominalism (e.g., Sober 1980), but Nanay (2010) argues that it does.

If Darwin is a nominalist about species, then it stands to reason that he would adopt this view of emotion types as well. Sure enough, in both

[27] Charles Bessey embraced this view in the early twentieth century, writing, "Nature produces individuals and nothing more.... Species have no actual existence in nature. They are mental concepts and nothing more.... Species have been invented in order that we may refer to great numbers of individuals collectively" (Bessey 1908; quoted in Mayr 1968, 163). Thanks to Andrea Scarantino for pointing me to this passage.

[28] Not all scholars agree that Darwin was a nominalist. See Ghiselin (1969), Beatty (1985), and Stamos (2007).

Descent and *Expression*, he claims that he has classified emotions according to convenience, mirroring his claims from *Origin* that species are classified according to convenience:

> As no classification of the mental powers has been universally accepted, I shall arrange my remarks in the order most convenient for my purpose (1871, I:35)

> My observations will be arranged according to the order which I have found the most convenient; and this will generally lead to opposite emotions and sensations succeeding each other. (1872, 147)

Recall that most associationist philosophers present emotions in pairs of opposites. Hobbes discusses hope then despair, fear then courage (2020, VI.14–18, p. 44). Locke shifts from joy and sorrow to hope and fear (1894, II.20.7, p. 305). Hartley's order is love, then hate, then desire, then aversion (1749, I.369). Brown begins with happiness and melancholy (1824, II.52, p. 257), while Spencer begins with (offensive) fear and (defensive) anger (1855, §200). Darwin's first pairing flips the order of Brown's. His chapter on "low spirits" (1872, 178) is followed by a chapter on "high spirits" (1872, 198). Darwin's ordering follows the associationist tradition, but he emphasizes that this arrangement is purely a matter of "convenience," echoing Buffon's claims about biological taxa and Brown's claims about emotion types.

Granted, Darwin's remarks about convenience refer to his order of emotions, as opposed to his types of emotions. But if we look at the types of emotions that he discusses, we will find that he views them as conventional groupings of emotions – as mental abstractions – that are neither categorical nor hierarchical. In *Expression*, he analyzes the expressions of more than fifty types of emotions, as shown in Table 4.2.

This is, I believe, the longest list of emotions found in eighteenth- or nineteenth-century emotions research. It is longer than Locke's, Hume's, and Hartley's lists combined.[29] Hobbes and Brown also proposed long lists, but Darwin's list is nearly twice as long as either of theirs.[30]

Throughout *Expression*, Darwin emphasizes that these types cannot be clearly distinguished from one another. Consider contempt (#11) and disgust (#18). If these terms pick out natural groupings of emotions, then the relation between them ought to be clear. Either one emotion is a subtype of the other, or they are discrete and non-overlapping types, or

[29] See §§2.3–2.5 for Locke's, Hume's, and Hartley's lists.
[30] See §2.2 for Hobbes's list and §4.2 for Brown's list.

Table 4.2 *Darwin's emotions*

1. Admiration	20. Fear	39. Pride
2. Affirmation	21. Grief	40. Rage
3. Ambition	22. Guilt	41. Reflection
4. Anger	23. Hatred	42. Revenge
5. Anxiety	24. Helplessness	43. Scorn
6. Astonishment	25. High spirits	44. Shame
7. Attention	26. Horror	45. Shyness
8. Avarice	27. Humility	46. Slyness
9. Conceit	28. Ill-temper	47. Snarling
10. Confusion	29. Impotence	48. Sneering
11. Contempt	30. Indignation	49. Suffering
12. Defiance	31. Jealousy	50. Sulkiness
13. Dejection	32. Joy	51. Surprise
14. Determination	33. Loathing	52. Suspicion
15. Despair	34. Love	53. Sympathy
16. Devotion	35. Low spirits	54. Tender feelings
17. Disdain	36. Meditation	55. Terror
18. Disgust	37. Modesty	56. Vanity
19. Envy	38. Negation	

one is a compound emotion, made up of the other type plus something else. Darwin argues, to the contrary, that contempt and disgust cannot be clearly distinguished from one another, or for that matter from scorn (#43), disdain (#17), sneering (#48), defiance (#12), or loathing (#33).

> Scorn and disdain can hardly be distinguished from contempt, excepting that they imply a rather more angry frame of mind. Nor can they be clearly distinguished from the feelings discussed in the last chapter under the terms of sneering and defiance. Disgust is a sensation rather more distinct in its nature, and refers to something revolting, primarily in relation to the sense of taste, as actually perceived or vividly imagined; and secondarily to anything which causes a similar feeling, through the sense of smell, touch, and even of eyesight. Nevertheless, extreme contempt, or as it is often called loathing contempt, hardly differs from disgust. These several conditions of the mind are, therefore, nearly related; and each of them may be exhibited in many different ways. (Darwin 1872, 254)

With other emotions, too, Darwin emphasizes that categories blur into each other. "Prolonged suffering . . . passes into low spirits, grief, dejection, and despair" (1872, 147). "Attention . . . graduates into surprise; and this into astonishment; and this into stupefied amazement" (1872, 278). In every case, he rejects the project of taxonomizing emotion types. His goal is rather to describe the range of expressions to which we affix these conventional labels.

Another reason to think that Darwin sides with Brown is that Darwin considers, but then rejects, Spencer's basis for natural groupings of emotions. For Spencer, natural emotions are grouped according to biological function. What makes fear *fear* is that it serves to protect the animal from danger. What makes love *love* is that it serves to propagate the species. Natural emotions are, on Spencer's theory, primitive psychological adaptations.

Darwin denies that emotions are adaptations. He argues that they are, for the most part, purposeless products of association. Here I must tread carefully. It is generally accepted that Darwin views emotional *expressions* as purposeless products of association.[31] We smile, frown, and pout, not because these expressions serve any purpose, communicative or otherwise, but merely because they happen to be associated with specific feelings. On what basis can we extend this view of emotional expressions to the emotions more generally?

Darwin's reasoning about expressions applies equally to emotions. If emotions are trains of associations, as Darwin believes, then emotions and expressions are produced by the same underlying mechanism, that of association. Why do we frown when we are angry? Because past experiences forged an association between that feeling and that expression. Why do we experience anger in the first place? Because past experiences forged an association between that sensation, that thought, that feeling, and that expression. It's the same explanation. Darwin's thesis that we produce expressions from the force of habit, even when they are of the least use, is a corollary of his thesis that we experience emotions from the force of habit, even when they are of the least use. The ambition of the associationist project, of which Darwin is a part, is to explain as much of our psychological lives as possible with repeated cranks of the same association lever.

Let us look again at Darwin's three principles of expression, noting that they derive from general principles of association, which purport to explain the whole emotion. The first principle is based on contiguity. *Events that occur together in time become associated in the mind.* If an occurrence causes us to have sensations, and if we voluntary undertake specific actions, then associations may form. The next time we encounter a similar situation, we may anticipate the same sensations and involuntarily perform the same actions, "though they may not then be of the least use" (Darwin 1872, 28). To be clear, the associated sensations and expressions *can be* useful. If a foe

[31] I discussed this interpretation in §1.2. See Burkhardt (1985, 359), Richards (2009, 114), Fridlund (1992), and Campbell (1997a, 454). In §4.5, I discuss Darwin's (1872, 365–366) claim that expressions have *become* useful for communication.

attacks me, causing me to feel pain, and I voluntarily put my hands up to shield myself, then then next time I see a foe I may dread pain and involuntarily throw my hands up. In this case, the sensation and expression serve a purpose, since they hasten my defensive stance. But Darwin's point is that the underlying mechanism, association, is indifferent as to whether the sensation and expression provide a service or not. Often, my feelings of dread only detract from my well-being. Often, my instinctive cowering achieves nothing. My emotion is an automatic reenactment of past events through mind- and muscle-memory.

The second principle is based on contrast. *Opposing inputs, opposing outputs.* If seeing a foe leads me to anticipate pain and to retreat, then seeing the opposite of a foe, namely, a friend, will lead me to anticipate the opposite of pain, namely, pleasure, and to do the opposite of retreat, namely, approach. The principle of antithesis seeks to explain not only why we express joy by opening our arms for an embrace, but also why we experience joy upon seeing an old friend. Again, these sensations and expressions *can be* useful, but they don't have to be, and in any case the underlying mechanism of association is indifferent. Darwin emphasizes that the opposing associations will follow, "though these are of no use" (1872, 28).

The third principle accounts for malfunctioning associations. When the flow of energy along connected nerve cells is excessive or interrupted, energy may "jump" to other, nearby nerve cells. For instance, a train of thoughts connecting "sleep," "bed," "stars," and "sky" may be interrupted by a loud sound and then veer in a new direction: "book," "chair," "sit," and "lap." There may be no association between "sky" and "book," yet these thoughts follow simply because energy has jumped between adjacent nerve cells. Applied to emotion, Darwin argues that "when the sensorium is strongly excited," or when "the supply of nerve-force may, as it appears, be interrupted," then a train of sensations and expression can occur, which are not properly associated in the mind, but which are enacted by adjacent nerve pathways (1872, 29). Any benefit provided would be entirely accidental.

I have argued that Darwin's principles of expression reveal that emotions, and not merely their outward expressions, are the purposeless products of association. Thus, Darwin rejects Spencer's basis for positing natural groupings of emotion. He denies that emotions are defined by their biological functions. Instead, following Brown, he holds that our emotion words pick out conventional groupings of emotions, whose members vary widely in feeling, expression, and function. There is no essence of fear, anger, contempt, or loathing. Our emotion categories are vague and porous, bleeding into each other.

One could object to my claim that Darwin is a conventionalist and nominalist about emotion types by pointing to his observation that there are universal expressions of emotion. After all, Darwin writes that people everywhere express astonishment "by the eyes and mouth being opened wide, and by the eyebrows raised" (1872, 15), and that people everywhere express contempt "by a slight protrusion of the lips and by turning up the nose, with a slight expiration" (1872, 16). These claims seem to imply that all astonishments and all contempts can be naturally grouped together, based on their shared universal expressions.

In fact, Darwin observes that the emotions we group together under a common name are expressed in a wide variety of ways. Astonishment can be expressed by an opening of the mouth (Darwin 1872, 282) or by a protruding of the lips (1872, 285). Most people remain silent when astonished, but some people groan and others hiss or whistle 1872, 282–285). The hands express astonishment when thrown up in the air, but also when covering the mouth (1872, 288). Darwin writes of contempt that "it is natural that [it] should be expressed in many different ways, inasmuch as various habitual actions serve equally well, through the principle of association, for their expression" (1872, 254–255). He describes a range of movements around the nose and mouth that commonly express contempt (1872, 255–256). Smiling, laughing, spitting, and sticking out one's tongue are all common expressions of contempt (1872, 255, 260–261).

The emotions we group together as "astonishment" and "contempt" are expressed in a variety of ways. Darwin indeed claims that many of these expressions are universal, but what he means is only that each variant can be observed in people everywhere. It's not only Europeans who sometimes turn up their noses when experiencing an emotion we call "contempt"; Americans, Asians, Africans, and Australians sometimes turn up their noses, too. But people everywhere express their contempt in myriad other ways as well.[32]

Each emotion can be expressed in multiple behaviors. But each behavior can express multiple emotions, too. The behaviors that express astonishment can also express disappointment (Darwin 1872, 363), contempt (1872, 260), and helplessness (1872, 264). The behaviors that express

[32] Darwin notes that different cultures have adopted different hand gestures for indicating that someone is contemptible (1872, 257), for instance.

contempt can also express amusement (1872, 255), disgust (1872, 255, 260–261), and hatred (1872, 261). There is no strict correspondence between emotions and expressions, and therefore expressions do not provide a natural basis for grouping emotions together.

Darwin's view that there are universal expressions of emotion is consistent with his conventionalist, nominalist theory of emotion types. A person experiences a train of associations ending with raised eyebrows and an open mouth. We see this expression and we give the emotion a name, such as "astonishment." In doing so, we group this episode of emotion together with other episodes, based on a family resemblance. They don't all involve the same thoughts, feelings, or expressions, but they have enough in common to make the category convenient. Raised eyebrows and open mouths are universal expressions of astonishment insofar as this grouping includes episodes of emotion from every city and village around the world, which are likewise expressed in raised eyebrows and open mouths. We can also say that these behaviors are universal expressions of surprise, amazement, or awe, since these groupings overlap substantially with, without being identical to, the astonishment grouping.

In *Expression*, Darwin describes an informal experiment. He presented twenty-four people with a photograph of a man with eyebrows raised and mouth open, and asked them to name the emotion expressed. The photograph, taken from Duchenne's collection, is titled "Surprise." Darwin reports: "one alone did not at all understand what was intended. A second person answered terror, which is not far wrong; some of the others, however, added to the words surprise or astonishment, the epithets horrified, woeful, painful, or disgusted" (1872, 279). Darwin judges that this face is a universal expression of emotion – but which emotion, exactly? He accepts multiple correct answers: "surprise," "astonishment," "woeful astonishment," "fearful astonishment," "surprise & horror," "pain & wonder," and "stupid wonder."[33] One answer, "terror," he judges to be wrong, "but not far wrong." These names denote overlapping groupings of emotion, all of which include episodes expressed with eyebrows raised and mouth open.

Darwin's associationist theory seeks to explain why people everywhere express astonishment with raised eyebrows and open mouths, without appealing to an essence of astonishment. In each case, a sensation gives rise to an associated feeling, which gives rise to an associated thought,

[33] Darwin drew tables summarizing the results (DAR LIB 186, 27–29).

which gives rise to these associated expressions. At no point does Darwin's explanation require us to classify the emotion according to type.

I have argued in this section that Darwin sides with Brown in the Brown–Spencer debate. Darwin is a nominalist about species, and he takes the same approach toward emotions. Furthermore, Darwin considers but then rejects Spencer's basis for positing natural groupings of emotions. Finally, I considered an objection based on Darwin's claims about universal expressions of emotion but argued that these claims do not pose a problem for my interpretation.

At first glance, Darwin seems unreflective about emotion types. He writes confidently about the varying expressions of contempt, disgust, scorn, disdain, sneering, and loathing, without clarifying the differences between them. And he readily ascribes the same emotions to humans and nonhuman animals. If Darwin were a naturalist about emotion types, then we could rightfully criticize him for playing fast and loose with these classifications. The differences between contempt, scorn, disdain, and so on would need to be clearly stated. And evidence that nonhuman animals possess the essential properties of each type would need to be given. But if Darwin is a conventionalist, as I have argued, then the lack of rigor in his classification of emotions reflects a carefully considered philosophical position.

4.5 Basic Emotions

Emotions research has evolved since Darwin's *Expression*. The debate between Spencer's naturalism and Brown's conventionalism has matured into a debate between basic emotions theory (BET) and psychological constructionism (PC). Darwin has not been forgotten, however. Ekman (2009) claims that BET is the modern descendent of Darwin's *Expression*. Barrett (2013) replies that if you follow Darwin's good ideas in *Origin* and ignore his bad ideas in *Expression*, then you end up with PC. I shall argue that both sides in this contemporary debate misinterpret Darwin. It is not Darwin, but Spencer, who is the forefather of BET. And we needn't ignore *Expression* to see the connection between Darwin and PC, since *Expression* already anticipates PC's central tenets.

BET is the view that some emotions evolved as solutions to common survival challenges.[34] Fear evolved to help animals escape danger. Anger

[34] Philosophical treatments of basic emotions theory and the view that emotions are natural kinds include Griffiths (1997), Charland (2002), Prinz (2004), Green (2007), Scarantino & Griffiths (2011), Scarantino (2015), Glazer (2018a), and Hutto et al. (2018), among others. Scientific

evolved to help animals fend off competitors. Sadness evolved to help animals cope with loss. Because the basic emotions evolved long ago in a common ancestor of humans and other animals, these emotions are now experienced by most humans and many nonhuman animals.

According to BET, there is a relatively small number of basic emotion types that exist in nature.[35] Several lists have been proposed, yet most identify fewer than ten basic emotions. Robert Plutchik's (1962) list includes joy, sadness, anger, fear, surprise, disgust, anticipation, and trust. Paul Ekman's (1984) list shares the first six – joy, sadness, anger, fear, surprise, and disgust – but adds contempt and shame.[36] Jaak Panksepp's (1992) list shares three emotions with the first two – FEAR, ANGER, and SADNESS – but then adds SEEKING, LUST, CARE, and PLAY (see Table 4.3). (Panksepp writes emotions in all caps to remind readers that these terms correspond imperfectly to our common understanding of them.) Some advocates of BET, including Keltner et al. (2019), are open to there being many more natural kinds of emotion.

These short lists of basic emotions resemble Spencer's theory. Recall that Spencer posited just four basic emotions: fear, anger, aesthetic appreciation, and love. Fear and anger appear on everyone's lists of basic emotions. Spencer's aesthetic appreciation is similar to Plutchik's surprise and anticipation, Ekman's surprise, and Panksepp's SEEKING. Spencer's love is similar to Panksepp's LUST and Plutchik's and Ekman's joy. By contrast, Darwin discussed more than fifty emotions, without specifying any as the most basic.

formulations of basic emotions theory include Plutchik (1962), Ekman (1992), Panksepp (1992) and Keltner et al. (2019).

[35] Are basic emotions the only emotions? Or are there also non-basic emotions? Debate on this question continues. Ekman (1994) argues that there are no non-basic emotions – that every genuine emotion is instance of a basic emotion. Plutchik (2001) argues that basic emotions can combine to create non-basic emotions. He classifies contempt, for instance, as the combination of anger and disgust. Griffiths (1997) argues that beyond the basic emotions are "higher cognitive emotions" (such as guilt) and "socially constructed emotions" (such as *amok*, from the Malay word denoting a dissociative psychological disturbance causing a sudden and ferocious spree of attacks). Prinz (2004) argues that the names "fear" and "anger" denote non-basic, culturally informed cousins of unnamed basic emotions. Prinz (2012) elsewhere argues that there are a few basic emotions and many derived non-basic emotions.

[36] Ekman's list has grown over the years. Ekman (1994) splits joy into several distinct positive emotions, all of which share the smile as their typical expression: amusement, contentment, excitement, pride in achievement, relief, satisfaction, and sensory pleasure. Ekman also adds awe, embarrassment, guilt, and interest to his list, noting that they may not have distinctive expressions. But Ekman has also eliminated surprise from his list, doubting that it is really an emotion. So while Ekman (1984) names eight basic emotions, Ekman (1994) names seventeen.

Table 4.3 *Lists of basic emotions*

Plutchik (1962)	**Ekman (1984)**	**Panksepp (1992)**
Sadness	Sadness	SADNESS
Anger	Anger	ANGER
Fear	Fear	FEAR
Surprise	Surprise	SEEKING
Disgust	Disgust	CARE
Joy	Joy	LUST
Anticipation	Contempt	PLAY
Trust	Shame	

Why does BET propose such short lists? Why these emotions as opposed to others? BET holds that an emotion must meet a set of conditions to count as a basic emotion. First, it must "have relevance to basic biological adaptive processes" (Plutchik 1962, 68). In other words, it must solve a common survival challenge. In his early formulation of basic emotions theory, Plutchik identified eight such challenges, corresponding to each of his hypothesized basic emotions:

1. *Incorporation.* The act of taking in or ingesting food ...
2. *Rejection.* ... behavior involved in getting rid of something harmful which has already been incorporated ... such as expelling feces or vomiting.
3. *Destruction.* This prototypic pattern of behavior occurs when the organism contacts a barrier to the satisfaction of some need, and consists essentially in an attempt to destroy the barrier ...
4. *Protection.* ... an attempt to avoid being destroyed ...
5. *Reproduction.* ... the prototypic response associated with sexual behavior ...
6. *Deprivation.* The loss of a pleasurable object which was been contacted or incorporated ...
7. *Orientation.* ... the pattern of behavior which occurs when an organism contacts a new or strange object ...
8. *Exploration.* ... the more-or-less random activities organisms use to explore their environment. (1962, 72–74)

Plutchik's theory resembles Spencer's, which likewise matches basic emotions with biological functions. Both use the term "destruction" to denote anger's function and "protection" to denote fear's function. Plutchik posits two functions related to exploration (recognizing the familiar, pursuing the unfamiliar), and Spencer similarly emphasized the interplay of familiarity and unfamiliarity in aesthetic appreciation. Finally, Plutchik and Spencer both identify love with "reproduction." By contrast, Darwin rejects the identification of emotions with biological functions. He claims that emotions are, in many cases, purposeless products of association.

A related feature of basic emotions is that they may have dedicated neural circuity.[37] Panksepp states that:

> 1. The underlying circuits are genetically prewired and designed to respond unconditionally to stimuli arising from major life-challenging circumstances . . .
> 2. The circuits organize diverse behaviors by activating or inhibiting motor subroutines . . . that have proved adaptive in the face of life-challenging circumstances during the evolutionary history of the species . . .
> 3. Emotive circuits change the sensitivities of sensory systems relevant for the behavior sequences that have been aroused. (Panksepp 1992, 557)

Spencer agrees, claiming that there are special ganglia corresponding to each basic emotion, which he calls the "seat of the emotion" (1855, §204). Darwin, by contrast, does not weigh in on the neurophysiology of emotion in *Expression*.[38] Since he thinks that emotion categories are conventional and tend to blur into each other, he should deny that emotion types are localized in the brain.

A third feature of basic emotions, according to BET, is that they tend to have universal behavioral expressions. Ekman focuses on facial expressions, arguing, "The strongest evidence for distinguishing one emotion from another comes from research on facial expressions. There is robust, consistent evidence of a distinctive, universal facial expression for anger, fear, [joy], sadness, and disgust" (Ekman 1992, 189). Nearly everyone, he claims, naturally expresses joy by smiling, sadness by pouting, anger by frowning, and so on. That being said, Ekman also posits culturally specific "display rules" to explain why some people suppress some expressions in some situations. An early study (Ekman et al. 1969) found that while American and Japanese subjects expressed negative emotions while watching movies alone in a room, only American subjects expressed negative emotions while watching the same movies in the company of a scientist. Ekman hypothesized that the Japanese subjects suppressed their

[37] For defenses of this view, see Panksepp (2008) and Panksepp and Watt (2011). For a critical evaluation of the evidence, see Lindquist et al. (2012) and Barrett (2017b).

[38] However, Richards (2005, 169–170) discusses Darwin's exploration of this topic in his early notebooks.

expressions in order to comply with social norms. Thus, although people naturally express the same emotions in the same ways, there can be social pressures to voluntarily suppress these expressions. While Ekman focuses on voluntary modifications of expressions, others have argued that there are cultural differences in spontaneous expressions that resemble dialects (Elfenbein & Ambady 2002; Elfenbein et al. 2007). There is one shared language of emotional expressions, but many cultural dialects.

On this point, Spencer and Darwin agree. Darwin likewise acknowledges that socialization leads to the introduction of some new expressions and the unlearning of some old ones.[39] For instance, he claims that the clenching of fists in anger is restricted to cultures with a history of fist fighting (1872, 246–247). And he claims that "civilized" children unlearn the "savage" tendency to protrude the lips when sulking (1872, 234–235). Darwin's view seems to be that we inherit a first draft of emotional expression, which is revised and altered over the course of experience.

A fourth feature of basic emotions is that they are found in other animal species. Plutchik writes that basic emotions should "be found in some form at *all* evolutionary levels" (Plutchik 1962, 68). Ekman writes that basic emotions should have "comparable expressions in other animals," adding that "researchers have observed that other primates generate facial expressions similar to those observed in humans" (Ekman 1992, 178–179). Panksepp studied laughter in rats (Panksepp & Burgdorf 2003; Panksepp 2011).

Spencer and Darwin both agree. In the introduction to *Expression*, Darwin objects to Sir Charles Bell's claim that nonhuman animal expressions pale in comparison to human expressions. Darwin replies, "But man himself cannot express love and humility by external signs, so plainly as does a dog, when with drooping ears, hanging lips, flexuous body, and wagging tail, he meets his beloved master" (1872, 11). Two chapters of *Expression* are dedicated to the expressions of nonhuman animals, with chapter 4 exploring expressions common across taxa and chapter 5 detailing expressions unique to specific taxa.

In sum, Darwin's theory agrees with BET on a couple of points but disagrees on other, more fundamental points. Darwin affirms that there are universal expressions of emotion, despite cultural variations, and he affirms that some emotions and expressions are shared by humans and nonhuman animals. However, Darwin denies that there is a short list of basic emotions, grouped together naturally according to biological function and

[39] I discussed these points in §3.7.

localized in the brain. Spencer's theory agrees with BET on every point. He claims that there is a short list of adaptive, localized emotions, found in many different species, and expressed universally among humans. Not only that, but Spencer's list of emotions overlaps substantially with the lists proposed by Plutchik, Ekman, and Panksepp.

Advocates of BET claim to be walking in the footprints of Darwin, but it was in fact Spencer who forged their path.

4.6 Constructed Emotions

Constructionists deny that emotions like anger and fear are natural kinds. They claim that these emotions are "constructions." But what does that mean? It will take some work to flesh this out. To begin, we can distinguish two strains of constructionism.

Whereas BET takes a biological perspective on emotions, *social constructionism* (SC) takes a sociological perspective, theorizing that emotions are the products of social norms and practices. Fear, anger, and ennui emerged somewhat recently in human history, and only in particular cultures, once those cultures developed language games for these constructs. On this approach, emotions are presumed to exist only once people think they exist (Ratner 1989, 212). A hypothetical culture with no emotion concepts would have no emotions either (Briggs 1970). People would have feelings and physiological reactions, of course, but these feelings and reactions would not necessarily align with our familiar emotion scripts, nor would they have the same social significance.

By contrast, *psychological constructionism* (PC) takes a psychological approach to emotions, theorizing that emotions are the products of simpler psychological processes within individuals.[40] Anger is not hardwired in your brain, but you experience anger when your brain mixes together the ingredients of anger in the right way. Your anger may resemble your parents' and grandparents' anger, but it's not because you inherited a dedicated anger mechanism from them. Rather, each of you simply followed the same recipe to bake the same anger cake. Psychological constructionists disagree as to whether emotion concepts are among the necessary ingredients for emotions (Scarantino 2015, 347). Barrett (2006;

[40] Scientific formulations of psychological constructionism include Averill (1980a; 1980b), Russell (2003), and Barrett (2006; 2013; 2017a; 2017b). Some view Schachter & Singer (1962) as a precursor. Philosophical treatments of the social and/or psychological construction of emotion include Sartre (1948), Armon-Jones (1986), Harré (1986), and Huebner & Glazer (2016). See also Ratner (1989).

2012; 2017b) says yes: Anger occurs once a person applies the anger concept to make sense of their feelings and physiological reactions.[41] Russell (2003) says no: Anger is a familiar pattern of feelings and physiological reactions, which can occur even if no one applies the anger concept to it. However, the application of an emotion concept is necessary for the "meta-experience" of emotion, or the state of experiencing anger *as anger*.

Broadly speaking, BET emphasizes cross-cultural similarities in emotion, SC emphasizes cross-cultural differences in emotion, while PC emphasizes the differences in emotion from person to person, within and across cultures.

Let us now set SC aside and try to state more clearly the differences between BET and PC. As we shall see, it is actually quite difficult to locate the border between them.

According to Gendron and Barrett (2009, 317) a defining feature of PC is its claim that "emotions are psychical compounds that are constructed out of more basic psychological ingredients that are not themselves specific to emotion." The idea is that while both BET and PC view emotions as compounds, BET posits *emotion-specific* components, while PC posits *emotion-independent* components. BET says that when you experience an emotion, say, anger, you experience feelings and perform behaviors that are unique to anger.[42] PC says that these feelings and behaviors can occur outside of anger, too.

Notice that Gendron and Barrett's definition of PC applies perfectly to all associationist theories of emotion. Associationists claim that emotions are aggregates of simpler states, which can and do occur independently of emotion, yet which are bound together by the forces of association. According to David Hartley, "our Passions or Affections can be no more than Aggregates of Simple Ideas united by Association" (1749, I.368). According to Alexander Bain:

> the theory of Emotion will consist in the statement of the general laws that regulate the concurrence, coalition, transfer, or aggregation of the primary feelings – the Muscular states and the Sensations; and the analysis of the several Emotions will be the account of the component feelings in each, and the mode of their composition under some of these general laws. (1865, 35–36)

[41] In more recent work, Barrett (2017b) explains the application of emotion concepts using a predictive coding framework. On the latter, see Clark (2013).

[42] Scarantino (2015) argues that BET *doesn't* say this. My point here is just that Gendron and Barrett claim that BET says this. I argue later that Gendron and Barrett's definition doesn't do BET justice.

And according to Herbert Spencer, love is made up of "a variety of pleasurable ideas, not in themselves amatory, but which have an organized relation to the amatory feeling" (1855, §202). In other words, none of these pleasurable ideas is specific to love, yet when their powers combine, they become love. An emotion is like a melody. Its notes occur in countless other melodies, but only if arranged or timed differently. When the notes occur in this order and with this timing, the melody is immediately recognizable.

Gendron and Barrett's definition of PC applies perfectly to Darwin's theory, too. Darwin theorizes that emotions are "trains of thought long associated with action" (DAR LIB 404, 333). These thoughts and actions necessarily occur independently of emotion, yet they become parts of emotions once they are linked by association: "ideas & bodily actions make the emotions" (Darwin 2009, N 89). This point is clearest in Darwin's analysis of emotional expressions.[43] Sir Charles Bell had claimed that there are special muscles in the face that exist solely for the purpose of expressing our emotions. (We could say that these muscles are *emotion-specific*.) Darwin refutes this claim by arguing that "[d]istinct uses, independently of expression, can indeed be assigned with much probability for almost all the facial muscles" (1872, 10). In other words, every muscle contraction that occurs in an emotional episode has a reason to occur in other circumstances as well. (We could say that the muscles are *emotion-independent*.) Darwin's first principle of expression, the principle of serviceable associated habits, asserts that the movements we call "emotional expressions" once occurred independently of emotion, but have become involuntary expressions "through the force of habit and association" (1872, 28). Once a particular thought or idea triggers an associated action, the whole episode has become an emotion.

I've argued that Darwin's theory satisfies Gendron and Barrett's definition of PC. However, this definition is outdated and no longer reflects the current state of the debate. The problem with the definition is that some versions of BET likewise analyze emotions as "psychical compounds that are constructed out of more basic psychological ingredients that are not themselves specific to emotion" (Gendron & Barrett 2009, 317). According to Andrea Scarantino's (2015, 345–346) version of BET, for instance, the basic emotions involve feelings, thoughts, desires, physiological changes, and facial expressions, all of which can occur independently of emotion. None of the elements is specific to emotion, yet there is

[43] This analysis is the subject of the next chapter.

nonetheless a special emotion mechanism that coordinates them. If the defining feature of PC is found also in BET, then it's not really a defining feature of PC.

The word "basic" is partly to blame for the confusion. Ortony and Turner (1990) point out that the "basic" in "basic emotions theory" can mean several different things. An emotion is *biologically basic* if it is connected to a basic adaptive process (Ortony & Turner 1990, 219–320). If fear and anger evolved to help us flee and fight, then fear and anger are biologically basic emotions. An emotion is *psychologically basic*, by contrast, if it cannot be broken down into simpler elements that occur independently of emotion (Ortony & Turner 1990, 324; Scarantino & Griffiths 2011).[44] If fear and anger involve feelings and behaviors that occur only in these emotions, then fear and anger are psychologically basic emotions, as I've defined them.

Gendron and Barrett seem to assume that BET is committed to psychologically basic emotions. However, some advocates of BET claim that it is committed only to biologically basic emotions, which may or may not be psychologically basic. We've seen that Scarantino's version of BET posits biologically basic emotions that are not psychologically basic, since they are made up of components that are not emotion-specific.

Let's try again to distinguish BET from PC. If it doesn't boil down to a disagreement over psychologically basic emotions, then perhaps it boils down to a disagreement over biologically basic emotions. Is it true that BET affirms biologically basic emotions while PC denies them?

Scarantino (2015, 346–347) argues that the fundamental difference between BET and PC is that the former accepts, while the latter rejects, emotion-specific mechanisms that elicit the cascade of components making up an emotion. Upon seeing a bear, your heart will race, your mouth will open, the tiny hairs on your body will stand on end, and your stomach will drop. BET and PC may agree on this order of events. However, BET claims that there is a special psychological entity or organ, alternately called an "affect program," "emotion system," or "emotion

[44] I have modified this definition to better fit my dialectic. Ortony and Turner write that emotions are psychologically basic if they "do not have other emotions as constituents" (1990, 324). Scarantino and Griffiths distinguish between two kinds of psychological basic-ness: "A discrete emotion is psychologically basic in the *emotion primitive* sense . . . if it does not have other discrete emotions as constituents and it is not a species of another discrete emotion.. . . An affective state is psychologically basic in the *affective primitive* sense . . . if it does not have other affective states as constituents and it is not a species of another affective state" (2011, 451). My definition suggests a third sense of psychological basic-ness, which more closely resembles Scarantino and Griffiths's second sense.

mechanism," dedicated to rapidly detecting threats and initiating defensive maneuvers (Scarantino 2015, 340–341). The sight of the bear, or perhaps the recognition that the bear poses a threat, activates the fear program/system/mechanism, which sends signals to the circulatory system, causing the heart to race; to the respiratory system, causing breathing to stop; to the integumentary system, causing body hairs to stand up; and to the digestive system, causing . . . I'd rather not say. On this view, the various components of emotion have a common psychological cause, the underlying program/system/mechanism. Griffiths (1997) argues that these psychological entities are natural kinds, which evolved precisely because they solved common survival challenges.

PC denies the existence of such emotion programs/systems/mechanisms. Yes, upon seeing a bear, you will experience a cascade of psychological, physiological, and behavioral changes. But these components are not controlled by a single emotion-specific mechanism. They are instead controlled by other domain-general mechanisms. According to Russell, the interactions of these diverse mechanisms nonetheless form recognizable patterns, which we are accustomed to calling "fear," "anger," "sadness," and the like (2003, 152). Barrett (2017b) adds that the brain rapidly forms predictions about these patterns based on implicit representations of emotion types.

Coan and Gonzales agree with Scarantino's account of the difference between BET and PC, but they describe it in slightly different terms. They say that BET analyzes emotions as *latent variables*, whereas PC analyzes emotions as *emergent variables* (Coan & Gonzalez 2015). Latent variables are the unobserved but inferred causes of observable events. If you see a person turn red in the face while shouting and slamming their fists, then you may infer that they are doing these things because they are angry. You don't see their anger, but you assume that anger is the unobserved cause of their observable behaviors. On this view, the emotion is the program/system/mechanism that computes sensory inputs into behavioral outputs.

If emotions are emergent variables, however, then they "do not cause, but rather *are caused by*, their measured indicators" (Coan & Gonzalez 2015, 213). This wording is not quite right, since it implies that the emotion is a temporally distinct event that is the causal effect of the measurable indicators. A better way to put it is that emotions "*emerge* from the interaction of more basic ingredients . . . imply[ing] that the product (the emotional instance as a whole) is more than the sum of its parts" (Barrett 2015, 64–65). An emotion is a pattern, a gestalt that appears in the interactions of elements. Anger happens when a person's face turns red while shouting and slamming their fists. Emotions are results.

On this account of the border between BET and PC, most associationist theories of emotion fall on the side of PC. These theories reject emotion-specific programs/systems/mechanisms. Instead, they maintain that the various components of emotion are controlled by domain-general mechanisms of association. Recall Hartley (1749, I.368) and Bain (1865, 35–36), quoted earlier. Bain elaborates:

> the theory of Emotion will consist in the statement of the *general laws* that regulate the concurrence, coalition, transfer, or aggregation of the primary feelings ... *Now, in addition to the adhesive property of Contiguous Association, two great principles operate in the concurrence of different feelings: the Law of Harmony and Conflict, and the Law of Relativity.* (1865, 35–36; italics added)

The appeal to "general laws" in accounting for emotional phenomena discounts the need for emotion-specific mechanisms. Upon seeing a bear, your heart races, your mouth opens, your body hairs stand on end, and your stomach drops – not because a fear program/system/mechanism has commanded it, but simply because circumstances have caused these events to co-occur in the past, linking them in your mind. Each event brings the next in tow.

Herbert Spencer is an interesting case to consider. On the one hand, he agrees with his fellow associationists that the components of emotion are combined via a general learning mechanism. He writes: "Everyone can testify that the psychical state which ... we call anger, consists of mental representations of the actions and impressions which would occur while inflicting some kind of pain on another" (1855, §200). Regularities in emotion mirror regularities in nature. No special emotional systems or mechanisms are needed.

But, on the other hand, he argues that emotions are inherited: "every emotion implies some portion of nervous structure by which its various elements are united" (1855, §204). Can we call this inherited structure a dedicated emotional system or an "affect program"? We can. Although this affect program was fashioned long ago by a general learning mechanism, it exists in us today as an acquired faculty of emotion.

Recall that Spencer's philosophy is a synthetic philosophy. It embraces associationism *and* physiognomy. Or in today's lingo, it embraces BET *and* PC. PC explains the phylogeny of emotion, while BET explains the ontogeny of emotion.

If Spencer's theory straddles the fence between BET and PC, Darwin's plants its feet firmly in PC territory. Darwin denies that emotions are connected to basic adaptive processes, theorizing that many emotional expressions serve no purpose whatsoever. They occur now because they occurred before, and because habits and associations are stubborn. As long as they don't actively harm us, natural selection will permit them to linger.

More to the point, Darwin is emphatic that "[e]motions are results" as opposed to causes (DAR LIB 404, 333). Embracing the counter-intuitiveness of this proposal, he asks, "[B]ut what first caused this bodily action, if the emotion was not first felt? without abrupt flush, acceleration of pulse, or rigidity of muscles, – man cannot be said to be *angry*. He may have pain or pleasure these are sensations" (2009, N, 66). Anger is not the cause of the abrupt flush, the acceleration of the pulse, or the rigidity of muscles; rather, the emotion is the result. "Anger" is the emergent variable, what emerges from the interactions of the elements, not a latent variable presumed to cause them.

For Darwin, as for Russell, we don't need emotion concepts to experience emotions. A mouse surely experiences fear, even though it may not be able to conceptualize its experience as such. That being said, it's not as if there is a specific number of emotions that exist in nature, awaiting to be discovered. Rather, there are plural ways of carving up the patterns we find, some more useful than others. At times we may be content to name just four emotions; at others, we may lose count at fifty. Emotion terms pick out episodes that we group together conventionally for the sake of convenience.

In Darwin's view, a genuine science of emotion seeks to tell us not how many emotions there are, or how they may be taxonomized, but how the emotions are formed. Its attention ought to shift from emotions to their components – and especially to their expressions. So I turn now from Darwin's theory of emotions to his theory of expressions.

CHAPTER 5

Darwin's Expressions

5.1 The Bell–Darwin Debate

Darwin did not title his book *The Emotions in Man and Animals*. He titled it *The Expression of the Emotions in Man and Animals*. Philosophical ideas about the nature of emotion may have guided and constrained his investigation, but Darwin's explicit aim is to develop a theory of emotional expression – of the "outward manifestations" or "language of the emotions" (1872, 366, 367).

Perhaps this is why John Deigh writes that "Darwin himself was unconcerned with the question of the nature of emotions.... Darwin was chiefly concerned with involuntary expressions of emotion" (2004, 14–15). Even if the first half of this claim is mistaken (as I have argued throughout this book), the second half is surely correct. *Expression* is a book about, well, expression.

A standard line on *Expression* is that Darwin wrote it to challenge the creationist views of Sir Charles Bell.[1] Bell speculated that God created the human face with special muscles designed to reveal our innermost feelings (1844, 120). These muscles – and the expressions they conjure – are not found in any other species and serve no other function than to convey our uniquely human emotions.[2] Darwin replied that these muscles evolved in prehuman ancestors for reasons that have nothing to do with revealing our feelings (1872, 10). These muscles – and the expressions they conjure – are indeed found in many other species and serve other, more basic biological

[1] Browne (1985), Burkhardt (1985), and Richards (2005) emphasize this aim, as do Fridlund (1992) and Hartley (2001).

[2] These include the corrugator supercillii (a muscle above the eye) and the depressor anguli oris (a muscle around the mouth), both of which are involved in frowning (Bell 1824, 71–72; cited in Loudon 1982, 1795). Berkowitz (2015, 26) quotes a letter from Bell to his brother bragging of his discovery of these uniquely human muscles.

functions (or no functions at all).[3] According to this standard line, the Bell–Darwin debate was an empirical debate over the origins and functions of expressions. It was a debate between special creation and evolution – between religion and science. For this reason, the Bell–Darwin debate is sometimes presented in textbooks as the birth of the modern science of emotion.[4]

What scholars have overlooked about the Bell–Darwin debate, I argue, is that it was also a philosophical debate over the concept of emotional expression. Bell and Darwin disagreed over what the words mean. The former defined "expressing" as the conveying of emotion *by design* (Bell 1844, 120), while the latter defined "expressing" as the conveying of emotion *by accident* (Darwin 1872, 10).[5] Bell's concept is teleological: Emotional expressions are for communication. Darwin's concept is etiological: Emotional expressions are not for communication, but knowledgeable observers may nonetheless see an expression and infer its underlying causes. Is an emotional expression a message intended for its audience, or is it a reflex that unintentionally reveals a secret? This is the philosophical dimension of the Bell–Darwin debate (see Table 5.1).

A disagreement over definitions may seem trivial. However, it becomes important once we place the Bell–Darwin debate in its historical context. When Bell wrote his essays on emotional expression, Erasmus Darwin was still the talk of the scientific town.[6] Recall that Erasmus was Charles's grandfather, and that besides proposing an early theory of evolution, Erasmus developed an associationist theory of emotion. According to Erasmus, expressions convey emotions only by accident, never by design. Thus, when Bell argued that expressions convey emotions by design, we should read this as a rejection of the *idée du jour* – of Erasmus Darwin's

[3] Darwin writes of the corrugator supercillii that "both the orang and chimpanzee are said to possess this muscle, but it seems rarely brought into action, at least in a conspicuous manner" (1872, 143). However, he then reports his own observations of chimpanzees using this muscle to frown. Darwin does not mention the depressor anguli oris occurring in nonhuman animals.

[4] For instance, Workman and Reader's *Evolutionary Psychology* (2014) begins the chapter on emotion with a section entitled "Darwin, James and Freud and the Early Study of Emotion" (2014, 331).

[5] I have simplified the terms of the debate for the sake of introducing it. In addition to disagreeing over whether expressions convey emotions by design or by accidence, Bell and Darwin disagreed over how observers "read" these emotions – by sensing them or by inferring them.

[6] King-Hele sums up Erasmus's influence on British intellectual society at the turn of the nineteenth century as follows: "Darwin was eminent as a physician, and even more as a medical author; an imaginative mechanical inventor; a respected man of science, an innovator in both physical and biological science; a man of great social skill and energy; a founder of societies; the leading English poet of the 1790s, who profoundly influenced the Romantic poets" (2005, 19). However, "Darwin's reputation, so high in the 1790s, declined throughout the nineteenth century, rapidly and first, and later more slowly, to a nadir around 1900" (King-Hele 2005, 22).

Table 5.1 *Dimensions of the Bell–Darwin debate*

	Charles Bell	Charles Darwin
Empirical claim about the *origins* of expressions	God designed the human face with special muscles of expression	Humans inherited the muscles of expression from prehuman ancestors
Empirical claim about the *functions* of expressions	Smiles and frowns occur for the sake of conveying emotions (and for no other reasons)	Smiles and frowns occur for other biological reasons (and not for the sake of conveying emotions)
Philosophical claim about the *concept* of expression	To "express" an emotion means to convey that emotion by design	To "express" an emotion means to convey that emotion by accident

associationist analysis.[7] Bell hoped to make a name for himself by challenging the biggest theory around.[8] Similarly, when Charles Darwin argued that expressions convey emotions by accident, we should read this as a defense of his grandfather's associationist theory. We know that Charles first developed his theory amid his reading of Erasmus's *Zoonomia*, and that he only later had the idea of publishing *Expression* as a refutation of Bell.[9]

One of my aims in this chapter is to uncover the neglected philosophical dimension of the Bell–Darwin debate, especially as it concerns the conceptual analysis of emotional expression, and to show that this debate has shaped the philosophy of expression for more than 150 years. Contemporary philosophers are still debating Bell's and Darwin's definitions, often without realizing it.

My other aim is to show that Darwin's definition of expression is inextricably tied to his associationism. His reasons for thinking that expressions convey emotions by accident flow from his belief that emotions are trains of feeling and action, linked by associations. We express our emotions as we do, not because these expressions were designed to

[7] Hughes and Gardner-Thorpe (2022, 207) observe that Bell "fail[ed] to take into account" Erasmus's theory. I argue that Bell is familiar with Erasmus's theory and is actively arguing against it.

[8] Berkowitz (2015, 25–30) documents Bell's ambitions in writing his essays on expression. He hoped to establish himself as an important natural philosopher and to be appointed to the Royal Academy of Arts. I wish to add that Bell began his essays before moving from Edinburgh to London, just after his former classmate Thomas Brown achieved recognition by writing a critique of Erasmus Darwin's theories. It stands to reason that Bell modeled his efforts on Brown's.

[9] I defend these claims in §5.6, below.

convey emotions, but because these expressions happened to occur alongside particular feelings in the past and are now connected to them by the force of (inherited) habit. The view that expressions are happy accidents is a quintessentially associationist view.

The chapter proceeds as follows. I begin by reviewing Darwin's descriptions of emotional expressions. These descriptions will help us to fix in our minds what exactly a definition of "expression" ought to capture. Afterward, I will present a framework for classifying different definitions of "expression." From there, I will work through Erasmus Darwin's, Charles Bell's, and Charles Darwin's definitions, examining how each philosopher responded to those who came before him. Finally, I will document how the philosophy of expression has evolved – or rather, regressed – since the Bell–Darwin debate. I will demonstrate that philosophers have increasingly abandoned Darwin's definition for Bell's, perhaps without sufficient justification.

5.2 Charles Darwin's Descriptions of Expressions

Before we can get into the different definitions of "emotional expression," we should have a better idea of the myriad behaviors included in this concept. Smiles, frowns, and pouts may be the first expressions that spring to mind. But what other sorts of behaviors and reactions are included?

Among *Expression*'s most lasting contributions are Darwin's rich and detailed descriptions of emotional expressions. Darwin was, above all else, a naturalist. He had a sharp eye for detail and an immense appreciation of the complexity of small, seemingly insignificant things. Darwin's descriptions will alert us to the vast range of behaviors and physiological reactions that Darwin includes in his concept of "emotional expression."

The following are three representative descriptions:

1. When a dog approaches a strange dog or man in a savage or hostile frame of mind he walks upright and very stiffly; his head is slightly raised, or not much lowered; the tail is held erect and quite rigid; the hairs bristle, especially along the neck and back; the pricked ears are directed forwards, and the eyes have a fixed stare.... As he prepares to spring with a savage growl on his enemy, the canine teeth are uncovered, and the ears are pressed close backwards on the head.... Let us now suppose that the dog suddenly discovers that the man whom he is approaching, is not a stranger, but his master; and let it be observed how completely and instantaneously his whole bearing is

reversed. Instead of walking upright, the body sinks downwards or even crouches, and is thrown into flexuous movements; his tail, instead of being held stiff and upright, is lowered and wagged from side to side; his hair instantly becomes smooth; his ears are depressed and drawn backwards, but not closely to the head; and his lips hang loosely. From the drawing back of the ears, the eyelids become elongated, and the eyes no longer appear round and staring. (Darwin 1872, 50–51)

2. Birds belonging to all the chief Orders ruffle their feathers when angry or frightened. Every one must have seen two cocks, even quite young birds, preparing to fight with erected neck-hackles; nor can these feathers when erected serve as a means of defence, for cock-fighters have found by experience that it is advantageous to trim them. The male Ruff (*Machetes pugnax*) likewise erects its collar of feathers when fighting. When a dog approaches a common hen with her chickens, she spreads out her wings, raises her tail, ruffles all her feathers, and looking as ferocious as possible, dashes at the intruder. The tail is not always held in exactly the same position; it is sometimes so much erected, that the central feathers, as in the accompanying drawing, almost touch the back. Swans, when angered, likewise raise their wings and tail, and erect their feathers. They open their beaks, and make by paddling little rapid starts forwards, against any one who approaches the water's edge too closely. (1872, 97–98)
3. Under moderate anger the action of the heart is a little increased, the colour heightened, and the eyes become bright. The respiration is likewise a little hurried; and as all the muscles serving for this function act in association, the wings of the nostrils are somewhat raised to allow of a free indraught of air; and this is a highly characteristic sign of indignation. The mouth is commonly compressed, and there is almost always a frown on the brow. Instead of the frantic gestures of extreme rage, an indignant man unconsciously throws himself into an attitude ready for attacking or striking his enemy, whom he will perhaps scan from head to foot in defiance. He carries his head erect, with his chest well expanded, and the feet planted firmly on the ground. He holds his arms in various positions, with one or both elbows squared, or with the arms rigidly suspended by his sides. With Europeans the fists are commonly clenched.... Rage, anger, and indignation are exhibited in nearly the same manner throughout the world.... There is, however, an exception with respect to clenching the fists, which seems confined chiefly to the men who fight with their

> fists. With the Australians only one of my informants has seen the fists clenched.. . . Another observer speaks of the native men, when enraged, throwing their arms wildly about. (1872, 246–247)

These passages show that Darwin's descriptions are much more nuanced, and much more varied, than some have claimed.

Paul Ekman attributes to Darwin the view that emotions appear "primarily on the face" (2009, 3449). Ekman's own research focuses on facial expressions. But Darwin, as we have just seen, identifies expressive movements throughout the body.[10] Dogs don't just curl their lips and pull back their ears when angry. They also lower their heads, extend their tails, and bristle. Swans don't just open their beaks. They raise their wings and tails and extend their feathers. Humans don't just flare their nostrils and frown. They puff out their chests, plant their feet, and clench their fists or throw their arms about.

Ekman focuses on the face, and even more specifically on the muscles of the face. In his view, a facial expression is a coordinated set of muscular contractions. He developed the Facial Action Coding System (FACS) to break facial expressions down into their components (Ekman & Friesen 1978). Each muscle in the face – or more specifically, each visible movement of one or more muscles in the face – is assigned a number. For instance, the contraction of the frontalis (pars medialis), which results in the raising of the inner brow, is assigned the number 1. Using this system, Ekman codes a characteristic facial expression of anger as 4 + 5 + 7 + 23 (Ekman & Friesen 1978) (see Table 5.2).

Ekman notes that Darwin "did not attempt to provide a method for measuring facial movement" but that "FACS provides the means for describing all the variations [of them]" (2009, 3451).

Darwin looks beyond the face and beyond the muscular system.[11] Indeed, he finds that multiple organ systems are involved in the expression of emotion. Human expressions of anger often involve changes to the circulatory, respiratory, and digestive systems. "The heart and circulation are always affected; the face reddens or becomes purple, with the veins on the forehead and neck distended" (Darwin 1872, 240). "The respiration is likewise affected; the chest heaves, and the dilated nostrils quiver" (1872, 240). "If there be much and rapid speaking, the mouth froths" (1872,

[10] To be fair, Ekman acknowledges that Darwin "did give some attention to vocalizations, tears and posture" (2009, 3449), but even this clarification undersells Darwin's view that expressions mobilize the entire body.

[11] William James (1884, 191) noticed this as well.

Table 5.2 *FACS analysis of a standard facial expression of anger*

#	FACS name	Muscular basis
4	Brow lowerer	Depressor glabellae Depressor supercilii Corrugator supercilii
5	Upper lid raiser	Levator palpebrae superioris Superior tarsal muscle
7	Lid tightener	Orbicularis oculi (pars palpebralis)
23	Lip tightener	Orbicularis oris

242). In detailing the expressions of joy and amusement, Darwin observes that "the respiration is much disturbed ... the head and face become engorged with blood, with the veins distended ... Tears are freely shed" (1872, 208).[12] Grief likewise involves changes to multiple organ systems (1872, 178–179). Expressions mobilize the whole body. Not only the face, and not only the muscular system.[13] (See Table 5.3.)

To be fair, Ekman's own research highlights the characteristic physiological changes that occur in different emotions, but he contrasts these changes with the expressions that appear primarily on the face (1992, 175–183). Facial expressions are one thing, physiological changes another. The former occur for the sake of communicating emotions to observers; the latter do not. Darwin instead views all these events as equally expressive. The face may draw and hold our attention, but it is not a privileged site of expression.

Ekman also attributes to Darwin the view that emotional expressions are universal (2009, 3450). This is mostly right. As I argued in §3.7, Darwin claims that many expressions are universal, but he is also keen to emphasize cultural differences when he finds them. In a passage quoted earlier, he argues that the clenching of the fists is an expression of anger found only in societies practicing fisticuffs. Individuals raised in societies that practice other martial arts may express anger by flailing their open hands.[14]

[12] Darwin also relates an unusual example of a person who laughed until he vomited (1872, 76n10).

[13] William James rightly observed that Darwin documents "internal" as well as "external" expressions of emotion but felt that Darwin didn't go far enough: "But not even a Darwin has exhaustively enumerated *all* of the bodily affections characteristic of any one of the standard emotions" (James 1884, 191).

[14] Ekman notes that Darwin "proposed that gestures are culture-specific conventions" (2009, 3450). He means gestures like the thumbs-up or A-OK sign, which are not expressions of emotion but social signals. He doesn't mean emotionally expressive gestures like the clenching of fists.

Table 5.3 *Darwin's descriptions of expressions*

Organ system	Anger	Joy	Grief
Muscular	"[T]rembling is a frequent consequence of extreme rage" (1872, 241–242)	"Joy, when intense, leads to various purposeless movements – to dancing about, clapping the hands, stamping, &c., and to loud laughter" (1872, 198)	"The corners of the mouth are drawn downwards" (1872, 179)
Circulatory	"The heart and circulation are always affected; the face reddens or becomes purple, with the veins on the forehead and neck distended" (1872, 240)	"[T]he head and face become engorged with blood, with the veins distended" (1872, 208)	"The circulation becomes languid; the face pale" (1872, 178)
Respiratory	"The respiration is likewise affected; the chest heaves, and the dilated nostrils quiver" (1872, 240)	"[T]he respiration is much disturbed" (1872, 208)	"The breathing becomes slow and feeble, and is often interrupted by deep sighs" (1872, 179)
Digestive	"If there be much and rapid speaking, the mouth froths" (1872, 242)		
Integumentary		"Tears are freely shed" (1872, 208)	"[T]he eyes become dull and lack expression, and are often slightly suffused with tears" (1872, 179)

Of crying while laughing, Darwin notes that this habit is found in "most of the races of men," but that "in North America, the same fact has been observed in . . . [one] isolated tribe, but chiefly with the women; in another tribe it was observed only on a single occasion" (1872, 209–210). Darwin did not insist that crying while laughing is a universal expression. Darwin's view seems to be that many expressions are based on a universal template – so much so that these expressions are not plausibly conventional or learned – yet cultural differences emerge and persist.[15]

Ekman claims that Darwin focuses on universal facial expressions. But it was Darwin's creationist adversaries who privileged universal facial expressions.[16] Bell wished to direct our attention away from the sorts of physiological changes that occur in many species and toward the special muscles of the human face, which he claimed were put in us by God for the sole purpose of revealing our feelings (1844, 121). Trembling, gasping, and heaving do not deserve the honor of being called "expressions." Only smiling, frowning, and pouting are worthy of the name. (Bell includes blushing also.[17]) Darwin responds by placing the physiological changes and facial expressions on the same level. All are (or can be) equally expressive.

Darwin's strategy aligns with his associationist theory of emotion. He defines emotions as "trains of thought long associated with action" (DAR LIB 404, 333). When actions become associated with feelings, such that the onset of feeling triggers action, emotions form. It does not matter whether these actions are facial muscle contractions or changes to the respiratory, circulatory, or digestive systems. Whatever the action may be, as long as it is part of the cluster of actions associated with those emotional feelings, they qualify as emotional expressions. Crucially, these actions need not serve any particular purpose. They need not occur for the sake of conveying emotions. What makes them expressions is that they occurred at the right place at the right time. They occurred alongside a feeling, and by force of association and habit, they became linked in the mind to that feeling. Expressions do convey emotions, but they do so as happy accidents of association.

Bell and Darwin may have disagreed about whether physiological changes are expressions. But they agreed that spoken language is not a means of expression. Expression is inherently nonlinguistic. Bell famously quipped that "expression is to passion what language is to thought" (1844, 197). He adds, "As our thoughts are embodied and the reasoning powers developed by the instrument of speech, the passions or emotions have also a corresponding organ to give them a determined character and force"

[15] In addition to cultural variations in expression, Darwin was also interested in more idiosyncratic expressions. He mentions, for instance, "the loss of colour in the hair, which has occasionally been observed after extreme terror or grief" (1872, 67).

[16] Ekman (2009, 3450) recognizes that creationists also focused on universal facial expressions. Ekman and I differ in our assessment of whether or to what extent Darwin expanded his focus beyond universal facial expressions.

[17] See Darwin's (1872, 337) discussion of Bell's (1844, 95) remark.

(1844, 197). In Bell's dichotomy, spoken language is a conventional, nation- or region-specific means of communicating ideas, while expression is a natural, universal means of communicating emotions (1844, 53). An utterance of "I am sad" would not be an expression of sadness unless it were uttered with a pout or tear, and then it would be the pout or tear that expresses sadness.[18]

Darwin embraces this dichotomy. He writes, "The movements of expression give vividness and energy to our spoken words. They reveal the thoughts and intentions of others more truly than do words, which may be falsified" (1872, 366). Pouts and tears "give vividness and energy" to an utterance of "I am sad," but the latter by itself is not a means of emotional expression. Darwin adds another contrast to the language/expression dichotomy. Spoken language can easily deceive, while expression can be feigned or hidden only with skill and effort.[19] Relatedly, spoken language generally involves communicative intentions, while behavioral expressions generally do not. The latter normally occur spontaneously and independently of the will, which is what makes them more credible.

My goal in this section has been to clarify the range of behaviors that are included in the concept of "expression." We've seen that Bell has a somewhat narrow concept of expression, which includes only universal facial expressions (and blushing). Darwin has a much broader concept of expression, which includes also various changes in the circulatory, respiratory, digestive, and other organ systems. Both Bell and Darwin have narrower concepts of expression than some contemporary philosophers who also include language and art in their concept of expression.

Now that we have a better idea of what all may be included in the concept of emotional expression, we can attempt to define this concept. What does it *mean* to say that these behaviors "express" emotions? This is a distinctively philosophical question about concepts. I'll begin with a contemporary definition, which will introduce us to the terms we need to understand the Bell–Darwin debate.

5.3 Defining Expression

Mitchell Green is a contemporary American philosopher who writes on topics related to mind and language. His book, *Self-Expression* (2007), was the first monograph on expression to be published in nearly

[18] See Glazer (2017; 2023) for recent elaborations of this idea.

[19] Ekman (2009, 3451) notes that Darwin says surprisingly little to account for this contrast.

forty years.[20] He has updated and expanded upon his analysis in several subsequent publications (Green 2010a; 2010b; 2016; 2019).

Green's concept of "self-expression" is broader than Darwin's concept of "expression" in two respects. First, Green's concept includes the expression of all kinds of mental states. We express not only our emotions, but also our beliefs, intentions, and desires.[21] Darwin's concept, by contrast, includes only the expression of emotion. Second, Green's concept includes behavioral, linguistic, and artistic expressions. I can express sadness by crying, by telling others how sad I am, or by playing a sad song on my banjo.[22] Darwin's concept, by contrast, includes only the behavioral expression of emotion.

In what follows, I will present Green's definition of self-expression only as it applies to the behavioral expression of emotion. I will not consider the expression of other mental states (such as beliefs or desires), and I will not consider the linguistic or artistic expression of emotion. For the sake of convenience, I will use the general term "expression" to mean "the behavioral expression of emotion."

Green defines expressing as the intersection of *signaling* and *showing* (2007, 212). To express an emotion, a behavior must both signal and show that emotion. Each of these concepts requires further elaboration.

5.3.1 Signal versus Cue

A signal is a special kind of cue. A *cue* is "any feature of an entity that conveys information (including misinformation)" (Green 2007, 5). Where there's smoke, there's fire. Smoke can inform people of a fire, and therefore smoke is a cue of fire. Fires don't create smoke for the sake of telling anyone that there's fire, but people can nonetheless infer fire from smoke. Similarly, trackers follow the traces that animals leave behind as they move through an environment. Animals leave footprints where they step, impressions in the grass where they sleep, and broken twigs where they walk. Each of these traces conveys information to an expert tracker. Each of these traces is therefore a cue of the animal's movement. Once again, the

[20] Alan Tormey published *The Concept of Expression* in 1971. Of course, many other books have explored expression, but as a means to some further aim.

[21] I have argued elsewhere (Glazer 2023) that the expression of emotion involves a narrower sense of "expression," while the expression of belief involves a broader sense of "expression."

[22] I have argued elsewhere (Glazer 2017) that the behavioral expression of emotion involves a narrower sense of "expression," while the linguistic and artistic expression of emotion involves a broader sense of "expression."

animal may not leave these traces with the goal of helping the tracker to follow it. But these traces do nonetheless help the tracker to follow.

I have given examples of cues that do not occur for the sake of conveying information. Signals are cues that do occur for the sake of conveying information. Or as Green puts it, "A *signal* is any cue that was designed for its ability to convey the information that it does" (2007, 5).[23] If a dog urinates on a fire hydrant to mark its territory, then it is doing so for the sake of telling other dogs that this fire hydrant is taken. Like pawprints in the mud, a scent marking tells others that the animal has been there. Unlike pawprints in the mud, a scent marking is made *because* it tells others that the animal has been there. Communication is the scent marking's proper function, or what it evolved to do.[24] Similarly, if I light a beacon fire to alert someone at a distance that the barbarian hordes are approaching, then the resulting smoke isn't just a cue; it is also a signal. Smoke fills the air *because* it communicates an enemy invasion.

To say that a signal occurs for the sake of conveying information is to say that it does so "by design" (Green 2007, 5). This design can come in several forms. It may be the result of a convention, like the beacon fire. Or it may be the result of natural selection, like the scent marking. If an animal does something because its ancestor benefited from doing it, then we can say that the animal does it "by design." Conventions normally involve the intentions of intelligent agents. Natural selection normally does not.[25]

Cues and signals sometimes convey misinformation (Green 2007, 6).[26] When I unsuccessfully attempt to light a campfire, I produce lots of smoke without any fire. If someone at a nearby campsite were to see all the smoke and conclude that my campfire is lit, then they would be mistaken. Similarly, I might light a beacon fire to warn someone that the barbarian

[23] I have quoted Green's definitions of cues and signals, but these terms have been used by evolutionary ethologists for nearly a century. Konrad Lorenz (1939) is often credited with introducing the distinction between signals and cues. See Maynard Smith & Harper (2003) for a recent theory of cues and signals. See Funkhouser (2017; 2018) and Glazer (2018b) for a lively debate over whether signals and cues need to be perceptible, a debate I lost, sadly.

[24] I use the phrase "proper function" in Ruth Millikan's sense: "A function *F* is a direct proper function of *x* if *x* exists having a character *C* because by having *C* it *can* perform *F*" (1984, 26). More simply: *F* is a proper function of *x* if "*x* exists in order to perform *F*" (Millikan 1984, 26).

[25] Green writes: "When the process of natural selection accounts for a trait as an adaptation, that shows it to be designed without thereby justifying the inference that there must have been an intelligent design*er*" (2007, 5).

[26] Much has been written about the mechanisms that keep signals reliable. See Zahavi (1975; 1977), Maynard Smith (1976; 1978), and Zahavi & Zahavi (1997) for early discussions of the handicap principle, or the idea that signals are reliable if difficult or costly to fake. See Penn & Számadó (2020) for a recent critique of the handicap principle.

hordes are approaching, even if they aren't. I might be crying wolf. Speaking of which, anytime I take my dogs to a new park, they mark a territory that is not in fact theirs.

By this point, the distinction between cues and signals should be relatively clear. All signals are cues. But not all cues are signals. Signals are cues that communicate by design. Cues are things that communicate either by design (in the case of signaling cues) or by accident (in the case of non-signaling cues).

5.3.2 Show versus Indicate

I have explained the first term in Green's definition. To express an emotion, a behavior must signal an emotion. But this is not enough. An expression must also show an emotion.

In Green's analysis, to *show* an object is to enable perceptual knowledge of that object.[27] For instance, I can show you my badass Darwin tattoo by rolling up my sleeve. I can show you my dogs by pulling up a photograph of them on my phone. And I can show you a magic trick by performing it in your presence. In all three cases, I *show* so that you may *see*.[28]

Showing an object is different from indicating an object. To *indicate* an object is to enable someone to infer that object.[29] I can indicate my Darwin tattoo by telling you about it. I can indicate my dogs by showing you the scratch marks they left on my leg. (You see the scratch marks, but you infer the dogs that made them.) And I can indicate a magic trick by promising to show it to you later.

Showing an object enables perceptual knowledge. Indicating an object enables, at best, inferential knowledge. I say "at best" because it is possible to indicate objects that do not exist. Such indications do not enable any kind of knowledge. Suppose that we are walking in the woods and hear a rustling sound. "Bigfoot," I whisper. I have indicated Bigfoot as the source of the rustling, but my indication does not enable you to know that it was Bigfoot. Showing, by contrast, requires the existence of the object shown.

[27] Green (2007, 47–49) distinguishes among three types of showing, corresponding to three types of knowing. I can show you *that* something is true, which gives you *factual knowledge*. I can show you *how* to do something, which gives you *practical knowledge*. And I can show you *an object*, which gives you *perceptual knowledge*. This last type of showing, showing objects, is most relevant for understanding how expressions show emotions.

[28] Or hear, or feel, or smell, or taste. Showing provides perceptual knowledge, which isn't restricted to visual perception.

[29] My use of the term "indicate" comes from Davis (1988, 280). Taylor (1979) and Bar-On (2010), among others, likewise emphasize the contrast between enabling perception and enabling inference.

If the thing that I purport to show is not real, then I am not actually showing it. I only purport to show it. "These are Bigfoot's bones," I may say while gesturing toward a pile of bones. In this case, I have not shown you Bigfoot's bones, since these bones are not Bigfoot's. I have merely indicated Bigfoot's bones.

Many philosophers have argued that perception involves unconscious inference (Helmholtz 1867; Hanson 1958; Rock 1983). If this is true, then we may classify all showing as indicating, but not all indicating as showing. We may think of "showing" as the enabling of inference *operating within perception* and "indicating" as the enabling of inference *regardless of whether it operates within perception or outside it.* The classification creates a pleasing symmetry between the two parts of Green's definition. A signal is a special kind of cue, just as showing is a special kind of indicating.

Showing enables perceptual knowledge. Why "enables"? Because an act of showing may not actually result in knowledge. I can show you my Darwin tattoo, yet you may refuse to believe that it is real. You may insist that it is a temporary tattoo. Your skepticism does not count against me having "shown" you my tattoo. To say that showing "enables" knowledge is to say that showing offers knowledge to the witness, but it's up to the witness to accept that knowledge. Skeptics and cynics may refuse it.

To gain knowledge from an act of showing, the witness must be able to *recognize* the object, or to perceive it as such. Here I draw on the distinction between perceiving and perceiving as. We may look at the same duck–rabbit drawing, yet I may see it as a rabbit while you may see it as a duck (Wittgenstein 2006, II.11). We perceive the same object, but we perceive it as two different kinds of objects. Perceptual knowledge involves conceptualization, and the correct conceptualization at that.[30]

5.3.3 Expression

Green argues that expressing is signaling plus showing. A pout expresses sadness just in case it both signals and shows sadness. A guffaw expresses amusement just in case it both signals and shows amusement. If either condition is missing – if a behavior does not signal an emotion, or if it does not show it – then the behavior does not express the emotion. My aim in this subsection is to present Green's rationale for this analysis – why he thinks that expressing requires both signaling and showing. I'll start with the showing condition and work my way back to the signaling condition.

[30] However, see Orlandi (2011) for a compelling argument that seeing-as does not require concepts.

The idea that expressions show emotions flies in the face of common sense. Common sense says that we see an expression and must infer the underlying emotion.[31] I can see your pout, but I can't see your sadness. Your sadness is a subjective feeling that only you have direct access to. However, I can hypothesize that you are pouting because you are sad. I can infer your sadness from your pout. The commonsense view is that expressions indicate emotions. They do not – indeed cannot – show them.

Green is neither the first nor the last philosopher to challenge this commonsense view of expression. Ludwig Wittgenstein, Max Scheler, and Charles Taylor, among others, have likewise claimed that one can perceive emotions directly in their behavioral expressions:

> "We see emotion." – As opposed to what? – We do not see facial contortions and make the inference that he is feeling joy, grief, boredom. We describe a face immediately as sad, radiant, bored, even when we are unable to give any other description of the features. – Grief, one would like to say, is personified in the face. (Wittgenstein, 1980, §570)[32]

> [W]e certainly believe ourselves to be directly acquainted with another person's joy in his laughter, with his sorrow and pain in his tears, with his shame in his blushing, with his entreaty in his outstretched hands.... If anyone tells me that this is not "perception" . . . I would beg him to . . . address the phenomenological facts. (Scheler 1954, 260)[33]

> When I know something or something is plain to me, through an inference, there is something else which I know or which is plain to me in a more direct way, and which I recognize as grounding my inference.... It is characteristic of expression that it is not like this. I see the joy on your face, hear the sadness in the music. There is no set of properties that I notice from which I infer to your emotions or to the mood of the music. (Taylor 1979, 74)[34]

One might worry that the language of "perceiving" emotion is metaphorical. After all, Lakoff and Johnson (1980) identify "perceiving" as a common metaphor for "knowing," and thus the intuitive ring of "perceiving" emotions in their expressions may reflect only the idea that we can "know" emotions from their expressions.

31 See Stout (2010) for a recent defense of the commonsense view.

32 See Overgaard (2007) for an interpretation and defense of Wittgenstein's claim.

33 See Gallagher (2008), Krueger (2012), and Krueger & Overgaard (2012) for interpretations and defenses of Scheler's claim.

34 Glazer (2017) defends a perceptual analysis of expression inspired by Taylor's claim.

In defense of the idea that we can literally perceive emotions in their expressions, Green (2007, 86–89) offers the following argument:[35]

1. Expressions are parts of emotions.
2. Perceiving a part of an object is sufficient for perceiving that object.
3. Therefore, perceiving an expression is sufficient for perceiving an emotion.

The first premise of the argument is that emotions are not merely subjective mental states; they are complex psychological, physiological, and behavioral responses to common life challenges. Fear isn't just the sinking feeling in the pit of your stomach. It is also the hair standing on your neck and your jaw dropping. Fear is the whole sequence of events, which includes mental states, physiological reactions, and behavioral expressions. This first premise makes a substantive and controversial claim about the ontology of emotion. Although it is compelling, we may reasonably reject this premise.[36]

Green's second premise is that seeing a *part* of something is generally sufficient for seeing that *something*. Consider any ordinary object near you, like a closed book. You can see the book. But actually, you are seeing only a part of that book. You are seeing the part of its opaque surface that is facing you. You do not see the rest of its opaque surface. Nor do you see the inside of it. Still, if you've seen *part* of the book, then you've seen *the book*. Green calls this "part-whole perception" (2007, 86).

It follows from these two premises that if you've seen a *part* of an emotion, namely, the expressive component, then you've seen *the emotion*, even though there are, of course, other parts of it that you haven't seen and may never see. We see emotional expressions every day. Therefore, we see emotions every day, too.

This argument establishes that expressions enable the *perception* of emotion. It has not yet established that expressions enable *perceptual knowledge* of emotions. What's missing is recognition. A witness must be able to perceive an expression *as* an emotion (Green 2007, 89). No recognition, no knowledge.

Newen, Welpinghus, and Juckel (2015, 197–198) make a compelling case for the conclusion that emotion recognition is a form of pattern

[35] See Glazer (2018a) for a critical examination of this argument.

[36] Rowson (forthcoming) argues against the views that emotions are objects and that expressions are parts of emotions. However, she defends the conclusion that we can literally perceive emotions in their expressions by positing that expressions are the "stuff" of emotions, conceived as events as opposed to objects. I discuss her view further in §6.5.

recognition. They argue that we perceive an emotion when we perceive multiple features that have a high likelihood of co-occurring only as components of an emotion.[37] For instance, we may perceive that someone is snarling, clenching his fists, and turning red. People tend to do these things only when angry, and so we may perceive the person as angry. Newen et al. (2015, 199) argue that this process does not require an inference from the expressions to the emotion. Notice that this argument, like Green's argument, relies on the assumption that emotions are complex psychological, physiological, and behavioral states. If emotions are *just* feelings or cognitions, then both arguments fail.

So far, I have presented Green's argument for why expressing is showing. The following is Green's argument for why expressing is signaling, too:

> While self-expressions show one's thought, feeling, or experience, it is not the case that all phenomena that show such things are self-expressions.... A galvanic skin response might show a primate's fear, but does not express his fear. Likewise, consider the difference between blushing and weeping. A blush shows my embarrassment, but it doesn't seem natural to say that it expresses my embarrassment.... It is more natural to describe weeping as an expression than it is to describe blushing as an expression because we find it more natural to construe the former as a signal than the latter.
>
> We might of course be wrong. Advances in evolutionary biology of emotional expression might show, for instance, that blushing did evolve to signal embarrassment. In that case, consistent with the familiar phenomenon of common sense being educable, the intuitions to which I've just appealed would turn out mistaken. (2007, 27)

Green's argument appeals to linguistic intuitions. He asserts that "it doesn't seem natural to say that [a blush] expresses my embarrassment" and then explains this intuition by noting that blushing, in contrast to other expressions, does not signal emotion.

Not everyone shares Green's intuition. Darwin calls blushing "the most peculiar and the most human of all expressions" (1872, 310). Clearly, Darwin thinks that we should count blushing as an "expression" even if blushing does not signal embarrassment. Thus, while Green analyzes expressions as signals, we might prefer to analyze expressions as cues, which include those expressions that convey emotions by design *and* those that convey emotions by accident (like blushing).[38]

[37] Glazer (2023) defends a similar conclusion from slightly different theoretical assumptions.
[38] Scarantino (2019, 58) follows this path.

Table 5.4 *Definitions of expression*

	Signal	Cue
Show	Expressions convey emotions by design . . . . . . and enable perceptual knowledge of emotions.	Expressions convey emotions by design or by accident . . . . . . and enable perceptual knowledge of emotions.
Indicate	Expressions convey emotion by design . . . . . . and enable (at best) inferential knowledge of emotions.	Expressions convey emotions by design or by accident . . . . . . and enable (at best) inferential knowledge of emotions.

In sum, Green's arguments may be compelling, but they are not conclusive. He argues that expressions must show emotions, but one could reasonably defend the view that expressions need only indicate emotions. Similarly, he argues that expressions must signal emotions, but one could reasonably defend the view that expressions need only cue emotions.

Based on these considerations, we can distinguish four possible definitions of expression. These definitions differ across two axes: the signal-cue axis and the show-indicate axis (see Table 5.4).

I argue that this way of carving up the possibilities can be traced back to the Bell–Darwin debate. Bell argued that expressions signal and show emotions. Darwin argued that expressions cue and indicate emotions. For the past 150 years, philosophers of expression have rehashed the Bell–Darwin debate, often without realizing it. Some have argued that expressions are signals, others that expressions are cues. Some have argued that expressions show; others that expressions indicate.[39]

In what follows, I present an origin story for the contemporary philosophy of expression. I aim to show that Erasmus Darwin, Charles Bell, and Charles Darwin set the terms of the debate that rages on today.

[39] A position explored in the empirical literature, which I do not engage with here, is that smiles and frowns are indeed signals, but signals of *social intentions* as opposed to *emotions* (Hinde 1985; Fridlund 1994; Fernandez-Dols & Ruiz-Belda 1997; Fridlund 2017; Barrett et al. 2012; for critical perspectives, see Griffiths & Scarantino 2005; Green 2007; Scarantino 2017; Glazer 2019). A frown means that one is willing to fight. A smile means that one is willing to cooperate or make concessions. One could say that these behaviors are signals of social intentions and cues of emotion, if these expressions convey emotions by accident. Or one could say that these behaviors are neither signals nor cues of emotion, if they do not convey emotions at all. Such a person might deny that emotions are real or that they are expressed in the ways we take them to be. Some who have denied that smiles and frowns are signals of emotion have also denied that smiles and frowns are "expressions" of emotion (Hinde 1985; Fernandez-Dols & Ruiz-Belda 1997). Notice that this appears to assume that expressing necessarily involves signaling, a claim that Darwin rejects.

5.4 Erasmus Darwin's Definition of Expression

The Bell–Darwin debate should really be called the Bell–*Darwins* debate. Bell developed his definition in reaction to Erasmus Darwin's definition, and Charles Darwin defended Erasmus's definition from Bell. The Bell–Darwins debate thus begins with the elder Darwin. I presented Erasmus Darwin's theory of emotion in §3.2. Here I will repeat only the parts that are most relevant to the present discussion.

Erasmus claims that an emotion is a train of feelings and actions, linked by associations. In the traumatic moment of birth, for instance, an infant feels distress and gasps for air (E. Darwin 1796, 16.8.1). An association forms. As a result, when this individual feels a similar distress, even many years later, they may habitually gasp for air (E. Darwin 1796, 16.8.1). The general principle is that expressions are behaviors that happened to coincide with feelings in the past, and subsequently became linked to those feelings by the power of association. Now, anytime the feeling occurs, the expression is sure to follow. *Events that occur together in time become linked in the mind.* Erasmus emphasizes that these expressive habits are, by and large, purposeless (1796, 16.8.1). Gasping for air served a purpose on its very first occasion – it helped the infant transition to a new environment – but once this behavior became an emotional expression, occurring automatically upon the feeling of distress, it ceased to serve any function.

Erasmus does not use the language of "signals" and "cues," but he makes it very clear that expressions do not typically occur for the sake of conveying emotions to observers. If they do convey emotions, then they do so accidentally. Communication is a useful side effect of the association between feeling and action, but communication is not its raison d'être. Expressions are therefore cues but not signals. This analysis is befitting of Erasmus's Enlightenment project, which is to take phenomena that appear to be the result of intelligent design, and to show that they are the predictable results of simple mechanical processes. Expressions may seem to be God's stamp on our faces, but they are in fact byproducts of the unyielding law of association.

Erasmus does not use the language of "showing" or "indicating" either, but he makes it clear that observers must infer emotions from emotional expressions. They cannot be sensed directly. He writes:

> There are two ways by which we become acquainted with the passions of others: first, by having observed the effects of them, as of fear or anger, on our own bodies, we know at sight when others are under the influence of these affections.... Secondly, when we put ourselves into the attitude that any passion naturally occasions, we soon in some degree acquire that passion. (1796, 16.7)

In this incredibly rich passage, Erasmus anticipates two recent theories of how we know other minds: *Theory Theory* (TT) and *Simulation Theory* (ST). According to TT, we infer others' emotions based on a folk theory of how people tend to behave when experiencing emotions (Morton 1980; Carruthers 1996). We presume that smiling people are happy, that crying people are sad, and that frowning people are angry. These associations are built up from personal experience but also from media consumption. Pictures, books, and videos present us with emotional archetypes that we apply in our daily lives.

According to ST, we infer others' emotions by putting ourselves in their shoes and by imagining what we would feel if we were them (Gordon 1986; Goldman 1989). If I were smiling ear-to-ear, then I'd be happy. If I were sobbing with tears running down my cheeks, then I'd be sad. If I were frowning and clenching my fists and teeth, then I'd be angry.

In contemporary debates, TT and ST are viewed as alternatives to *Direct Perception Theory* (DPT), or the idea that we can perceive emotions directly in people's expressions (Gallagher & Zahavi 2012). This debate maps onto the distinction between indicating and showing. TT and ST hold that expressions indicate emotions, since they enable inferential knowledge of emotions. DPT holds that expressions show emotions, since they enable perceptual knowledge of emotions.

Erasmus was not familiar with TT, ST, or DPT, of course, but the question in his day was whether our knowledge of others' emotions is grounded in learned associations (inference) or a special, innate sense (perception). The philosopher Thomas Reid put it like this in an essay originally published in 1785:[40]

> That many operations of the mind have their natural signs in the countenance, voice, and gesture, I suppose every man will admit.... The only question is, whether we understand the signification of those signs, by the constitution of our nature, by a kind of natural perception similar to the perceptions of sense; or whether we gradually learn the signification of such signs from experience, as we learn that smoke is a sign of fire, or that the freezing of water is a sign of cold? I take the first to be the truth. (Reid 1827, 310)

Reid affirms the view that we can perceive emotions by means of a special sense. Erasmus affirms the opposite view that we must infer emotions by means of learned associations.

Erasmus might have pursued the idea that expressions show emotions. After all, he embraces an ontology of emotion according to which

[40] See Maas (2003) for a helpful overview of Reid's philosophy of mind.

expressions are parts of emotions. Emotions are trains of feeling and action. We feel distress, we gasp for air, and the whole sequence constitutes fear. For this reason, emotions are not, strictly speaking, the causes of their expressions. Rather, one part of an emotion (the feeling) causes another part of that emotion (the expression).

Given this ontology of emotion, Erasmus might have affirmed that in seeing a part of an emotion (the expression), observers can thereby see the emotion. But even if he were willing to accept that expressions enable the *perception* of emotion, he would deny that expressions enable the *perceptual knowledge* of emotion. As an empiricist associationist, Erasmus wanted to posit the simplest of perceptual capacities possible and to have association – inference – do the work of getting a viewer from the perception of a bare particular to the conceptual knowledge of what kind of thing it is. For an associationist like Erasmus, knowledge of others' minds has to be inferential knowledge.

In sum, Erasmus Darwin defines expressing as cueing plus indicating. Bell responds by defining expressing as signaling plus showing.

5.5 Charles Bell's Definition of Expression

Sir Charles Bell (1774–1842) was an early neuro-anatomist and neurophysiologist. His greatest discovery was also his greatest controversy. In his privately circulated monograph, *An Idea of a New Anatomy of the Brain* (1811), he argued that there was a "grand division" between the functions of the cerebrum and the cerebellum (van Gijn 2011, 1190). He claimed that the cerebrum is responsible for sensation and motion, via the anterior nerve roots, whereas the cerebellum is responsible for visceral operations, via the posterior nerve roots. When François Magendie (1822) corrected Bell's theory, demonstrating that anterior nerve roots serve motor functions and posterior nerve roots serve sensory functions, Bell claimed credit for Magendie's discovery, allegedly falsifying documents to support his priority (Cranefield 1974). Today this division of nerve function is called the Bell–Magendie law.

Bell was also a gifted artist. He combined his talents by producing a series of beautiful illustrated books, entitled *Engravings of the Arteries* (1801), *The Anatomy of the Brain* (1802), *A Series of Engravings Explaining the Course of the Nerves* (1803), and *Illustrations of the Great Operations of Surgery* (1821).[41]

[41] See Neher (2008) for a brief discussion of Bell's relation to art.

Bell even dabbled in philosophy. As a university student, he studied under the philosopher Dugald Stewart (a vocal critic of associationism) and became captivated by William Paley's famous proof of God. Paley argues that God's existence can be inferred from human anatomy, especially the eye, which is too complex and sophisticated to be the product of mechanical laws. Paley's proof inspired Bell to write his *Essays on the Anatomy of Expression in Painting* (1806), later revised and republished as *Essays on the Anatomy and Philosophy of Expression* (1824, 1844).[42] In these essays, Bell argued that emotional expressions provide further evidence of God's intelligent design.[43] Like his other illustrated books, Bell's *Essays* contain many of his own illustrations.[44]

Of the *Essays*, the fifth is the most philosophical.[45] In it, Bell contrasts two definitions of expression and defends a more nuanced version of the first with a detailed discussion of human anatomy. The second definition is particularly interesting, since this is the definition that both Erasmus and Charles Darwin defend.

5.5.1 First Definition

The following is the first definition that Bell considers:

> The violent passions are exhibited so distinctly in the countenance of both man and animals, that we are led to consider the movements by which they are made obvious, as characteristic signs provided by nature for the express

[42] To be more precise, the first edition was titled *Essays on the Anatomy of Expression in Painting* (1806). The second edition was titled *Essays on the Anatomy and Philosophy of Expression* (1824). And the third was titled *The Anatomy and Philosophy of Expression as Connected with the Fine Arts* (1844). With each subsequent edition, Bell tinkered with the wordings of his points, but the substance remained the same. Observe, for instance, how this sentence evolved across the three editions:

1. "This view of things, however, so natural at first sight, is not altogether satisfactory to philosophy; and a more jealous observation of the facts seems to suggest an opposite theory" (1806, 84).
2. "This view, however, which appears to me so natural and just, is not received; an opposite theory has prevailed" (1824, 57).
3. "This view, however, so natural at first, is not altogether satisfactory; and an opposite theory has been proposed" (1844, 120).

I quote from the third edition (1844), since this is the edition that Darwin quotes from.

[43] See Berkowitz (2014) and Hughes & Gardner-Thorpe (2022) for discussions of Bell's relation to natural theology.

[44] Loudon (1982; 1795) notes that Bell also included illustrations by Joseph Saunders and David Wilkie without crediting them.

[45] In the first (1809) and third (1844) editions, this essay is the fifth in the collection. But in the second (1824), it is the fourth essay.

> purpose of intimating the inward emotions: that they may be interpreted by a peculiar and intuitive faculty in the observer. (1844, 120)

Because many expressions are so clear and distinct, it is natural to assume that they are "characteristic signs" that have "the express purpose of intimating the inward emotions." In other words, it is natural to assume that expressions are *signals*, which were designed for their ability to convey emotions to observers. Bell attributes this design to an intelligent designer, God, but one could also attribute this design to natural selection.

It is also natural to assume that expressions signal emotions by *showing* them. Bell writes that expressions are "interpreted by a peculiar and intuitive faculty in the observer." We have a special sense, dedicated to sensing emotions. We see a smile, and our joy-detector activates. We see a pout, and our sadness-detector activates. If you show an emotion to this special faculty, it will immediately register its presence. No inference is required. Recall that Thomas Reid had argued that we understand expressions "by a kind of natural perception similar to the perceptions of sense" (Reid 1827, 310). This is what Bell is affirming in his first definition.

In sum, Bell's first definition, like Mitchell Green's definition, asserts that expressing is signaling plus showing.

The problem with this definition, according to Bell, is that many nonhuman animal expressions are *not* signals. These expressions occur for reasons that have nothing to do with communication. Bell develops this objection amid his discussion of the second definition of expression.

5.5.2 *Second Definition*

The following is the second definition that Bell considers:

> [A]n opposite theory has been proposed, in which such special provision is denied, and the appearances are accounted for, as the effect of certain actions which are performed in obedience to the common laws of the animal economy. (1844, 120)

Bell writes that "special provision is denied," meaning that this definition denies that expressions are designed for their ability to convey emotions. Rather, they occur for reasons unrelated to communication. Bell gives several examples. A lion glares not to signal anger but "to see his prey more clearly" (1844, 121). Likewise, its snarl is "the natural act of unsheathing his fangs" (1844, 121). These expressions were designed for other purposes, related to "the animal economy," or to survival, plain and simple.

To be clear, these expressions *do* convey emotions to onlookers. If you see the lion's glare and hear its snarl, then you know that the lion is angry. According to the second definition, these expressions may not be signals, but they are certainly cues.

The second definition rejects the showing condition, too. It denies that we have a special faculty that enables us to perceive emotions in their expressions. Instead, it claims that we must infer emotions from their expressions, based on their past association:

> we are taught by experience alone, to distinguish the signs of the passion in man: that in infancy we learn that smiles are expressive of kindness, because accompanied by endearments, and that frowns are the reverse, because they are followed by blows. (1844, 120)

We infer happiness from smiles. Therefore, smiles indicate – but do not show show – happiness.[46]

In sum, the first definition states that expressing is signaling plus showing; the second definition states that expressing is cueing plus indicating. Bell does not say who proposed this second definition, but I suspect that he had Erasmus Darwin in mind. As I argued in §5.4, this was indeed Erasmus Darwin's definition. Not only this, but shortly before Bell wrote his essays, his former classmate Thomas Brown earned considerable acclaim by writing a critique of Erasmus Darwin's theory of mind and emotion (King-Hele 1999, 314–315). Bell wrote his essays in pursuit of similar fame, and Erasmus Darwin would have been an obvious target. Regardless of whether Bell had Erasmus Darwin in mind, Bell's second definition is the same as Erasmus Darwin's.

The problem with this second definition, according to Bell, is that it ignores important differences between human and nonhuman animal behavior. He develops this objection in the course of presenting his own view of expression, which offers something of a compromise between these two definitions.

5.5.3 Bell's View

Bell presents his own view of expression as follows:

> But, if we attend to the evidence of anatomical investigation, we shall perceive a remarkable difference between the provision for giving motion

[46] Reid anticipates this analysis and rejects it out of hand: "It seems to me incredible, that the notions men have of the expression of features, voice, and gesture are entirely the fruit of experience. Children, almost as soon as born, may be frightened and thrown into fits by a threatening or angry tone of voice" (1827, 310).

> to the features in animals, and that for bestowing expression in man. In the lower creatures, there is no expression, but what may be referred, more or less plainly, to their acts of volition, or necessary instincts; while in man there seems to be a special apparatus, for the purpose of enabling him to communicate with his fellow creatures, by that natural language, which is read in the changes to his countenance. (1844, 121)

At first glance, Bell's view appears to be a synthesis of the first two definitions. He agrees with the second definition that nonhuman animal expressions occur for reasons unrelated to communication yet can nonetheless inform curious onlookers of underlying emotions via association. Remember that the lion glares not to signal anger but to see its prey more clearly. And Bell agrees with the first definition that many *human* expressions occur for the sake of expressing emotions to curious onlookers who come equipped with special emotion-detectors. Humans smile for no other reason than to reveal their joy.

Were this a true synthesis, then Bell would conclude that humans and nonhuman animals have different ways of expressing their emotions. However, this is not what Bell concludes. Instead, he concludes that only humans "express" their emotions. Nonhuman animals reveal their emotions, but strictly speaking, "there is no expression, but what may be referred, more or less plainly, to their acts of volition, or necessary instincts" (1844, 121). In other words, Bell agrees with the first definition that expressing is, in every case, signaling plus showing. The first definition errs only in assuming that nonhuman animals signal and show their emotions. In fact, their behaviors do neither. Therefore, their behaviors are not actually "expressions" of emotion. The second definition describes nonhuman animal behaviors correctly yet errs in thinking that these behaviors are genuine "expressions" of emotion. See Table 5.5 for a summary of these different views on expression.

Table 5.5 *Bell's analysis of expression*

	First Definition	Second Definition	Bell's View
Do humans express emotions?	Yes	Yes	Yes
Do nonhuman animals express emotions?	Yes	Yes	No
Do expressions necessarily signal emotions?	Yes	No	Yes
Do expressions necessarily show emotions?	Yes	No	Yes

To defend his view, Bell appeals to "the evidence of anatomical investigation" (1844, 120). Many facial muscles can be assigned purposes unrelated to expression. But some, he claims, cannot. These muscles exist for the sake of expressing emotions to observers. And this clever design, he adds, is proof of an intelligent designer.

Bell's definition – that expressing is signaling plus showing – is identical to Green's definition. Moreover, notice how similar Bell's argument for the signaling condition is to Green's argument. Green acknowledges that galvanic skin responses and blushes are cues of emotions, since knowledgeable onlookers can infer emotions from them. But since these responses are not signals of emotions, they are not expressions of emotions either. To be an expression, a behavior must have been designed for the sake of revealing emotions. A key difference between Green's and Bell's views is Bell attributes this design to God, while Green attributes it to natural selection.

The problem with Bell's view, according to Charles Darwin, is that it ignores the descent of humans from nonhuman animals. What's true of nonhuman animal expressions is also true of human expressions. As we shall now see, Darwin offers a spirited defense of the second definition, which his grandfather had previously championed.

5.6 Charles Darwin's Definition of Expression

Charles Darwin published *The Expression of the Emotions in Man and Animals* in 1872. However, his theory of expression was decades in the making. He first sketched this theory in 1838–1839 in Notebooks M, N, and OUN. As I documented in §3.2, Darwin's early theorizing on expression was heavily influenced by his grandfather's *Zoonomia* (1796). Charles's first two principles of expression ("serviceable associated habits" and "antithesis") were inherited from Erasmus's theory.

These early notebooks on expression contain only a few passing references to Charles Bell. Notebooks M and OUN contain no references. Notebook N contains only three, and these amount to little more than Darwin dutifully copying Bell's descriptions of particular expressions (2009, M 41, 89, 103). This lack of engagement is somewhat surprising, given that Darwin had previously attended a meeting of the Plinian Society in which William Browne raised some of the same objections to Bell's theory that Darwin would later raise in *Expression* (Browne 1995, 77).

My hunch is that when Darwin first developed his theory of expression in 1838–1839, he was not thinking of the differences between his theory and Bell's. (It had been more than a decade since that 1826 meeting of the Plinian Society.) I suspect that Darwin's theorizing was guided primarily by his desire to update his grandfather's theory in light of his own theory of evolution. Indeed, the theory that emerges in those early notebooks requires no knowledge of Bell's theory (other than those particular descriptions that Darwin copied). Darwin could have come up with the same theory, even had Charles Bell never written his essays on expression.

Darwin's work on expression trailed off in the early 1840s but then picked up again in the late 1860s. In 1867, he sent his "Queries on Expression" to recipients around the world, requesting information on different peoples' expressions. In a letter to one of his recipients, he declares his intent to challenge Bell's analysis of expression:

> The subject [of emotional expression] is, I think, more curious and more amenable to scientific treatment than you seem willing to allow. I want, anyhow, to upset Sir C. Bell's view . . . that certain muscles have been given to man solely that he may reveal to other men his feelings. (Darwin 1867b; quoted in Burkhardt 1985, 358)

Darwin repeats this intent in the introduction to *Expression*:

> when I read Sir C. Bell's great work, his view, that man had been created with certain muscles specially adapted for the expression of his feelings, struck me as unsatisfactory . . . This belief led me to attempt the present work, however imperfectly it may have been executed. (1872, 19)

By this point, Darwin's framing of his theory has flipped. The 1838–1839 version of his theory makes many references to Erasmus and only a few to Bell. The 1867–1872 version makes many references to Bell and only a few to Erasmus. *Expression* cites Erasmus only three times, and these citations amount to little more than reports of Erasmus's observations of particular phenomena (1872, 30n3, 46n16, 77n11). There is no mention of Erasmus's principles of expression, which became Charles's principles of expression.

What led Charles to reconceive his theory as a refutation of Bell? We can only speculate. At that meeting of the Plinian Society in 1826, William Browne raised objections to the recently published second edition of Bell's *Essays* (1824). A third edition was published twenty years later in 1844, several years after Charles Darwin developed his 1838–1839 theory. We know that Darwin read this newer edition, since he quotes from it in *Expression*. Moreover, he remarks in a footnote that Bell's third edition is much superior to the first (1872, 2n4). I suspect that sometime after 1844 and before 1867 Darwin read Bell's third edition and was reminded of William Browne's objections.[47] He may not have remembered that he heard these objections from William Browne at that meeting of the Plinian Society in 1826, but the objections must have sprung into his mind. At this point, upon reflecting on his 1838–1839 theory of expression, he must have realized that this theory already contained a powerful argument against Bell. If expressions occur for reasons that have nothing to do with communication, and if these expressions are inherited from prehuman progenitors, then Bell's view that God equipped the human face with special muscles for expression is doubly wrong. In my speculative reconstruction of these events, Darwin must have been delighted to find, upon rereading Bell, that he had previously come up with a theory of expression that effectively undermined Bell's claims. When he writes in *Expression* that his reading of Bell "led me to attempt the present work," we can take this quite literally. He had already developed the theory of expression but hadn't published it. Once he realized that this theory could drive a nail in the coffin of one of his most prominent creationist adversaries, he found new motivation to publish.

Regardless of how it happened, Darwin did in fact frame *Expression* as a refutation of Bell. He challenges Bell's view that expressing is signaling plus showing by arguing that it is neither. Expressions don't have to signal emotions. And they don't have to show emotions either. For Charles, as for Erasmus, expressing is cueing plus indicating.

5.6.1 Against Signaling

Bell claims that expressing is, in every case, signaling. Darwin rejects this claim. He affirms that expressions convey emotions – that they are cues – but he denies that they were designed for this very purpose. Whereas Bell thinks of design in terms of special creation by an intelligent designer, Darwin thinks of design in terms of evolution.

Scholars generally agree that Darwin's goal in *Expression* is to demonstrate that expressions evolved for reasons that have nothing to do with communication.[48] We clench our fists in anger not for the sake of conveying anger but for the sake of preparing to fight (1872, 246–247). We weep in sadness not for the sake of conveying sadness but for the sake

[47] A fourth edition of Bell's *Essays* was published in 1847. I do not know when Charles Darwin acquired his copy of the third edition, nor do I know when he read it. The third edition was the only one found in Darwin's personal library.

[48] See Burkhardt (1985, 359), Fridlund (1992), and Campbell (1997a, 454).

of protecting the eyes from muscle strain (1872, 176). These expressions *do* communicate anger and sadness, but they do so accidentally, not by design (see 1872, 365–366). They are cues, but they are not signals.

We see Darwin's strategy play out in his three principles of expression. Through these principles, Darwin intends to show that "the habit of expressing our feelings by certain movements, though rendered innate, had been in some manner gradually acquired" (1872, 19). We habitually smile when happy. However, this habit is inherited. If we look back far enough, we will find an ancestor who did not smile instinctually, but who formed the habit of smiling when happy. And we will find that this habit was formed for reasons that have nothing to do with communication. In tracing the origins of our expressions, Darwin claims to have shown that expressions are not signals – that they were not originally designed for their ability to communicate emotions.

The first principle ("serviceable associated habits") states that when actions coincide with feelings, associations form, and later occurrences of those feelings will prompt the associated actions to follow (Darwin 1872, 28). If I am startled by a loud sound and widen my eyes to look for the source of the noise, then I may come to associate the feeling of being startled with the act of widening my eyes. As a result, I may automatically widen my eyes upon feeling startled, even if the source of my surprise is right in front of me and I have no need to widen my gaze. My child may be born with this inherited habit and involuntarily widen his eyes upon his first surprise. In this example, eye-widening is not a signal. It occurs for reasons that have nothing to do with communication. Yet it expresses my surprise nonetheless, since a knowledgeable witness can perceive my eye-widening and infer my feelings from them.

The second half of this principle takes Darwin's argument against signaling a step further. He argues that even when we attempt to hide our expressions, we perform actions that unintentionally reveal our feelings:

> Some actions ordinarily associated through habit with certain states of mind may be partially repressed through the will, and in such cases the muscles which are least under the separate control of the will are the most liable still to act, causing movements *which we recognize as expressive*. In certain other cases the checking of one habitual movement requires other slight movements; *and these are likewise expressive*. (1872, 28; italics added)

If I voluntarily squint in an attempt to hide the widening of my eyes when I am surprised, then this squinting, which has the "design" of hiding my surprise, may in fact express it. Darwin emphasizes here that behaviors are "expressive" in virtue of the meaning that observers make of them, not in virtue of their design. Behaviors are expressive when observers take them to convey emotions. For Charles, as for Erasmus, expressions are happy accidents of association and evolution.

In his presentation of the second principle ("antithesis"), Darwin again emphasizes that although contrasting expressions are not performed for the sake of revealing emotions, they may nonetheless reveal them:

> Certain states of the mind lead to certain habitual actions, which are of service, as under our first principle. Now when a directly opposite state of mind is induced, there is a strong and involuntary tendency to the performance of movements of a directly opposite nature, although these are of no use; and *such movements are in some cases highly expressive*. (1872, 28; italics added)

The movements "are of no use," but "are in some cases highly expressive." A knowledgeable observer can infer emotions from these movements, but the movements are not performed for this purpose.

The third principle ("direct action of the nervous system") follows suit. It identifies communication-independent reasons for expressions, and then affirms that they may communicate emotions as a fortuitous side effect:

> When the sensorium is strongly excited, nervous force is generated in excess, and is transmitted in certain definite directions, depending on the connection of the nerve-cells, and partly on habit: or the supply of nerve-force may, as it appears, be interrupted. Effects are thus produced *which we recognize as expressive*. (1872, 28–29; italics added)

In sum, a close reading of Darwin's principles reveals that Darwin's general aim is to show that expressions are not signals – that they do not communicate emotions by design – even though they are indeed cues – they communicate emotions by accident.

To be fair, Darwin concedes that *some* expressions may be preserved by natural selection insofar as they serve a role in communication, and that these expressions would be signals:

> On this principle [of antithesis] alone can I understand how the gestures and expressions which come under the present head of antithesis have originated. If indeed they are serviceable to man or to any other animal, in aid of inarticulate cries or language, they will likewise be voluntarily employed, and the habit will thus be strengthened. But whether or not of service as a means of communication, the tendency to perform opposite movements under opposite sensations or emotions would, if we may judge

> by analogy, become hereditary through long practice; and there cannot be a doubt that several expressive movements due to the principle of antithesis are inherited. (1872, 65)

Here, he insists that expressions come about for reasons that have nothing to do with communication. But he admits that some habits of expression may become "strengthened" if those expressions happen to convey emotions beneficially to observers. This would be an example of "exaptation" – of an adaptive trait taking on a new proper function (Green 2007, 115).

Keep in mind that a conceptual analysis seeks to identify necessary conditions. What does a behavior *have to do*, in order to qualify as an "expression"? Bell argues that behaviors have to signal emotions in order to be expressions, meaning that any behavior that fails to signal an emotion fails to express that emotion as well. Darwin has argued that behaviors don't have to signal emotions to express them, since most of the behaviors that we call "expressions" are not, in fact, signals. Darwin's argument remains strong, even if some behaviors are signals. Signaling is not necessary for expression. Only cueing is necessary. Since all signals are cues, behaviors that signal emotions will also satisfy this necessary condition for expressing.

5.6.2 Against Showing

Darwin also rejects Bell's claim that showing is a necessary condition for expressing. He argues that observers cannot perceive emotions directly, but have to infer emotions from their expressions, based on association.

Darwin develops this argument in the concluding chapter of *Expression*. He begins by noting that some nonhuman animals can learn how to read human expressions, but only after repeated interactions:

> Monkeys soon learn to distinguish, not only the tones of voice of their masters, but the expression of their faces, as is asserted by a careful observer. Dogs well know the difference between caressing and threatening gestures or tones; and they seem to recognise a compassionate tone.... This limited amount of knowledge has probably been gained, both by monkeys and dogs, through their associating harsh or kind treatment with our actions; and the knowledge certainly is not instinctive. (1872, 358)

He then adds that human children, too, learn how to read adult expressions based on association:

> Children, no doubt, would soon learn the movements of expression in their elders in the same manner as animals learn those of man. Moreover, when a

> child cries or laughs, he knows in a general manner what he is doing and what he feels; so that a very small exertion of reason would tell him what crying or laughing meant in others. (1872, 358)

I see you laugh, and I infer that you must be feeling what I feel when I laugh. The production and consumption of emotional expressions operate by the same underlying principle. In these cases, we learn to read emotions in the same way that we learn to say the ABCs or to play a game: by the power of association.

Having established that we can learn to associate emotions with expressions, Darwin then argues that we can inherit these associations from our progenitors. If we inherit the association, then we can instinctively infer emotions from their expressions. But Darwin is not saying that we have a special faculty of recognizing emotion. Rather, he is saying that we have some instinctual knowledge, based on the general principles of association and the inheritance of habit. Darwin writes:

> It is however extremely difficult to prove that our children instinctively recognise any expression. I attended to this point in my first-born infant, who could not have learnt anything by associating with other children, and I was convinced that he understood a smile and received pleasure from seeing one, answering it by another, at much too early an age to have learnt anything by experience.… When a few days over six months old, his nurse pretended to cry, and I saw that his face instantly assumed a melancholy expression, with the corners of the mouth strongly depressed; now this child could rarely have seen any other child crying, and never a grown-up person crying, and I should doubt whether at so early an age he could have reasoned on the subject. Therefore it seems to me that an innate feeling must have told him that the pretended crying of his nurse expressed grief; and this through the instinct of sympathy excited grief in him. (1872, 358–359)

The child saw the nurse crying, and the thought of grief popped into his mind. He would have no way of knowing why this thought popped into his mind. Darwin posits that his son had inherited an association between crying and grief, such that the sight of one automatically prompted the thought of the other. In this case, the inference is subconscious.

In sum, Darwin accounts for our ability to recognize expressions without positing a special sense, as Bell had done. Darwin's view is that of his grandfather's. The power of association can explain both why we perform specific movements in the throes of an emotion and why others can see our movements and know our feelings. Emotions are inferred from their expressions, based on their strong association in our minds. This association can be instinctual or learned.

Bell claims that expressions must both signal and show emotions. Darwin claims that neither signaling nor showing is required for expressing, and indeed that most expressions convey emotions by accident.

5.7 The Philosophy of Expression after Darwin

If we consider the Bell–Darwin debate as an empirical debate over the origins of expression, then Charles Darwin emerges as the clear winner. Evolutionary theories of expression have flourished. Creationist theories have floundered. But if we consider it as a philosophical debate over the concept of expression, then Bell has overtaken Darwin. Over the past 150 years, philosophers have steadily abandoned Darwin's camp and pitched their tent in Bell's. This is not to say that Darwin's camp is empty. I have identified one contemporary philosopher, Andrea Scarantino, who makes a compelling case for Darwin's definition.

The turn from cueing back to signaling began with the work of ethologists in the first half of the twentieth century. Oskar Heinroth (1911) and Julian Huxley (1914) argued that emotional expressions evolved as ritualized social signals. Dogs bare their teeth to communicate their willingness to bite. Birds flap their wings to communicate their willingness to fight. And gazelles jump in the air to communicate their ability to flee. These behaviors are exaggerated truncations of biting, fighting, and fleeing – exaggerated for the sake of effective communication. Whereas Darwin viewed expressions as cues, these ethologists viewed expressions as signals. By 1921, Wallace Craig (1921, 361) announced a consensus among ethologists that facial expressions are social signals, which evolved for the sake of communicating emotions to observers.

The turn from indicating back to showing began with the work of pragmatists and phenomenologists in this same period. John Dewey (1894; 1985) rejected the ontological distinction between emotion and expression, arguing instead that expressions are the perceptible components of complex psycho-physiological processes (Krueger 2014). For similar reasons, Ludwig Wittgenstein wrote that emotions are "personified in the face," by which he meant that we can see those emotions directly (Wittgenstein 1980, §570). Echoing Max Scheler's claim that we are "directly acquainted with another person's joy in his laughter" (Scheler 1954, 260), another phenomenologist, Maurice Merleau-Ponty, wrote that "I do not see anger or a threatening attitude as a psychic fact hidden behind the gesture, I read anger in it. The gesture *does not make me think* of anger, it is anger itself" (Merleau-Ponty 2002, 214; quoted in Krueger

2014, 145). These pragmatists and phenomenologists rejected the commonsense notion that we *infer* invisible emotions from visible expressions, espousing instead the view that we *perceive* emotions in their constitutive expressions. Expressions do not merely indicate emotions; they robustly show them.

In the second half of the twentieth century, analytic philosophers would try their hand at identifying the necessary and sufficient conditions for expression. One early effort by O. H. Green closely resembles Darwin's analysis of expressing as cueing plus indicating:

> It will be argued that a person's behavior, verbal or nonverbal, is an expression of emotion when and only when it (1) provides evidence for saying that the person presently has the emotion expressed, (2) occurs in the appropriate circumstances, (3) is subject to the person's control or modification, and (4) is honest. (Green 1970, 551)

I read his claim that an expression "provides evidence for . . . the emotion" as meaning that expressions provide inferential as opposed to perceptual knowledge of emotion. Expressions indicate; they do not show. Green adds that "behaviour which is an expression of emotion must be potentially teleological – either by way of [coping with an] emotion or an attempt to show others how one feels" (1970, 556). An expression *can* have the proper function of conveying emotion, but it *need* not. Some expressions occur for the sake of coping with emotions, such as when we slam our fists on the table to release our frustration. These "coping" expressions convey emotion by accident, not by design. Signaling is therefore not a necessary condition for expression, on Green's analysis. All expressions are cues, but not all expressions are signals. Although Green uses the word "show" in this passage, it functions merely as a synonym for "indicate." He doesn't mean that expressions provide perceptual knowledge of emotion.

One year later, Alan Tormey published *The Concept of Expression* (1971), the last monograph on expression until Mitchell Green's *Self-Expression* (2007). Tormey analyzes expression in a way that sounds at first like Darwin and O. H. Green:

> If *A*'s behavior *B* is an expression of *X*, then there is a warrantable inference from *B* to an intentional state of *A*, such that it would be true to say that *A* has (or is in state) *S*; and where *S* and *X* are identical. (1971, 43)

Although Tormey uses the word "inference," his explication of this relation hints toward the idea that expressions show, and do not merely indicate, emotions. He argues that "jealous behavior is not merely *evidence* for the presence of jealousy but, in an important sense, a constituent part of the complex referent of the predicate 'jealous'" (1971, 47). It's not that we infer an invisible emotion from a visible expression. Rather, the expression is the visible part of an emotion. The expression shows the emotion in the way that removing your hat shows your hair. Already we see movement away from Darwin's analysis and toward Bell's.

Charles Taylor makes the showing requirement even more explicit: "I *see* the joy on your face.... There is no set of properties that I notice from which I *infer* to your emotions" (Taylor 1979, 74; emphasis added). Taylor identifies showing as a necessary condition for expressing, but he does not identify signaling. As with Darwin and O. H. Green, expressions can show emotions by accident. They do not have to do so by design.[49]

With Tormey and Taylor, we see a shift from indicating to showing. We later see an independent shift from cueing to signaling. Wayne Davis claims that a behavior "is an expression of [an emotion] iff [it] spontaneously manifested [that emotion and] had no purpose beyond displaying [it]" (1988, 288; see also Davis 2008). Unlike O. H. Green, who made space for expressions that cope with emotions without displaying them, Davis insists that expressions must occur for no other reason than to display emotions. For Davis, "manifested" means the enabling of inference. The expression is evidence of the underlying emotion.

Rebecca Copenhaver and Jay Odenbaugh argue similarly that expressing requires signaling (and only signaling). They write that something expresses an emotion "if, and only if, it indicates the emotion and has the function of indicating it" (2020, 226).[50] By "has the function," they mean that this is a proper function, or that it occurs for the sake of performing this function. Copenhaver and Odenbaugh embrace the signaling requirement, but they reject the showing requirement as too strong. Expressing is signaling plus indicating.

We've seen that some contemporary philosophers have shifted from indicating to showing, while others have shifted from cueing to signaling. Still others have defended the view that expressing is showing and signaling. This was Bell's view. I already presented Mitchell Green's (2007) analysis. A similar analysis has been defended by Dorit Bar-On (2004).

[49] More recently, I have defended this analysis, according to which expressions show but need not signal emotions (Glazer 2017; 2023).

[50] I have quoted their definition of *artistic* expression, of how an artwork "expresses" an emotion, but they later clarify that this analysis applies also to behavioral expressions: "Embodied states express emotions because they have the function of indicating these emotions" (Copenhaver & Odenbaugh 2020, 227). Odenbaugh (n.d.) defends this view of behavioral expressions of emotion, too.

Bar-On suggests that behavioral expressions "are not merely designed to communicate information about the [emotion], but also serve to show the [emotion]" (2004, 271). For Bar-On, "showing" is the enabling of noninferential knowledge. This includes the enabling of perceptual knowledge, which is how I have previously defined "showing," but it also includes the enabling of other types of noninferential knowledge. Bar-On elaborates on this category by analyzing the sense in which linguistic utterances express emotions. Thus, Bar-On likewise analyzes expressing as signaling plus showing, though her sense of "showing" is broader than mine.[51]

Although many philosophers have left Darwin's camp for Bell's, the debate is not over (see Table 5.6). At least one philosopher, Andrea Scarantino, has recently argued against both signaling and showing requirements, leaving him with an analysis very close to Darwin's:

> *Natural emotional expression*: A sequence of bodily changes X of some agent A in context C *naturally expresses* A's emotion E relative to recipient R insofar as X increases the probability that A is undergoing E, relative to R's background knowledge. (Scarantino 2019, 51)

Much more simply, bodily changes express an emotion just in case they indicate that emotion. Indicating requires cueing but not signaling.

Scarantino (2019, 57) objects to the showing requirement by arguing that it construes expression too narrowly. Even if many expressions enable the perceptual knowledge of emotion, there are many other expressions that do not. I have been focusing on the behavioral expression of emotion – for example, smiles and frowns – but if we consider also the linguistic expression of emotion – for example, "I am glad," "I am mad" – then it becomes increasingly difficult to claim that all expressions enable the perceptual knowledge of emotion.[52] In hearing "I am glad," one does not thereby hear gladness. It seems more plausible to say that "I am glad" enables one to infer the speaker's gladness.

Scarantino raises two objections to the signaling requirement.[53] First, he points out that for many of the behaviors we call "expressions," we just

[51] Green's analysis of showing is broadest, since it allows for the enabling of any kind of knowledge. Bar-On's is narrower, since it excludes the enabling of inferential knowledge. Mine is the narrowest, since it excludes everything but the enabling of perceptual knowledge.

[52] Difficult, though perhaps not impossible. I have elsewhere (Glazer 2017) argued that even linguistic expressions must enable the perception of emotion to count as "expressions."

[53] In earlier work, Scarantino accepted the signaling requirement. For instance: "Emotional expressions *are* signals that represent, that manifest internal states, and that make demands on recipients *qua* signals" (2017, 171). However, in a 2019 paper, he revises his position, writing: "I reject the design requirement, even though in previous publications I accepted it for the sake of argument" (2019, 57).

Table 5.6 *Philosophers' definitions of expression*

	Signal	Cue
Show	Charles Bell	Max Scheler
	Dorit Bar-On	Ludwig Wittgenstein
	Mitchell Green	Charles Taylor
Indicate	Wayne Davis	Charles Darwin
	Rebecca Copenhaver	O. H. Green
	Jay Odenbaugh	Andrea Scarantino

don't know whether they are signals or cues. This "makes most of our claims about whether or not something is an emotional expression indeterminate" (2019, 57). Second, he observes that "whether a certain bodily change carries information *by design* or not makes no difference from the point of view of the communicative moves made available by the bodily change" (2019, 58). In other words, what matters is only that a behavior conveys emotion, not that it does so by design.

Scarantino's analysis is very similar to Darwin's. Expressions can be signals, but they don't have to be. Expressions can (and often do) convey emotions by accident. A subtle difference is that while Darwin insists that expressions always indicate emotions, Scarantino seems to allow for expressions to indicate or show emotions. If we broaden the definition of "indicating" to include the enabling of perceptual or inferential knowledge, such that the indicating/showing distinction parallels the cueing/signaling distinction as a genus–species relation, then Scarantino and Darwin would agree that indicating, but not showing, is a necessary condition for expressing.

Given Scarantino's arguments, it would be a mistake to conclude that Darwin's analysis is dead in the water. There is life to it yet.

CHAPTER 6

Darwin's Lost Legacy

6.1 Darwin's Forgotten Philosophy

I have argued that Darwin's *Expression* is a forgotten work of philosophy. Granted, philosophers cite and discuss this book with some frequency. But they do not engage with it as a work of philosophy.[1] Instead, they regard it as a purely scientific account of emotional expression, which sidesteps philosophical questions about the nature of emotion. John Deigh voices this view when he writes that "Darwin himself was unconcerned with the question of the nature of emotions.. . . Darwin was chiefly concerned with involuntary expressions of emotion" (2004, 14, 15). When philosophers of emotion cite Darwin's *Expression*, it is often for the sake of showing that their theories are consistent with Darwin's observations concerning the prevalence of certain expressions among humans and nonhuman animals.

I have argued, to the contrary, that Darwin's *Expression* is a work of associationist philosophy. Darwin's theory that emotions are "trains of thought long associated with action" (DAR LIB 404, 333) was first conceived in response to his grandfather's philosophy of emotion, later developed in dialogue with Hartley's philosophy of emotion, and ultimately refined in light of Spencer's philosophy of emotion. Darwin's theory is thoroughly and inescapably philosophical.

My aim in this final chapter is to explain why Darwin's philosophy of emotion was forgotten. The chief reason, I argue, is that associationist theories of emotion fell out of fashion very soon after Darwin published *Expression*, and philosophers and scientists were keen to forget Darwin's association with associationism. I trace this process of forgetting across three disciplines: biology, psychology, and philosophy. Concurrent

[1] A notable exception is Green (2007), who develops an analysis of expression in dialogue with Darwin. I analyzed the differences between Green's and Darwin's definitions of expression in §§5.3 and 5.6.

developments in these fields resulted in a collective amnesia regarding Darwin's philosophy of emotion.

In biology, the "Modern Synthesis" of Darwinian natural selection and Mendelian genetics drove the final nail in the Lamarckian theory that habits can be inherited. From this point on, scientists preferred to view Darwin's reliance on heritable habit in *Expression* as a puzzling mistake, which he ought to have avoided, as opposed to a deep and enduring commitment to an ill-fated theory of mind.

In psychology, associationism was dethroned by functionalism, which retroactively claimed Darwin as one of its own. William James and John Dewey presented Darwin as a precursor to their new theories, rather than a relic of older theories. Darwin's ties to associationism were subsequently severed.

Finally, twentieth-century philosophers construed the philosophy of emotion narrowly as the conceptual analysis of emotion. They left empirical questions concerning the formation and experience of emotion to the psychologists and busied themselves with specifying necessary and sufficient conditions for particular emotion types (anger, fear, disgust, etc.). Associationist theories placed great weight on speculative empirical claims, and comparatively little weight on conceptual claims, which contributed to philosophers' assessment that Darwin's *Expression* is not "philosophical" but rather "scientific."

Associationist theories of emotion were forgotten; Darwin's contributions to this tradition were also forgotten. But times have changed. In both philosophy and psychology, new views, which unwittingly resurrect aspects of Darwin's theory of emotion, have emerged. These new views demonstrate that Darwin's neglected theory ought to be remembered.

6.2 The Modern Synthesis

In *Evolution: The History of an Idea* (2003), Peter Bowler presents the standard narrative of the Modern Synthesis in biology. In the first quarter of the twentieth century, theories of adaptation and theories of genetics developed independently of one another. Adaptationists tended to embrace Lamarck's theory that adaptation occurs via the inheritance of acquired traits (Bowler 2003, 325). Geneticists marshalled evidence against Lamarck's theory but assumed that evolutionary change occurs on the basis of mutation, not adaptation (Bowler 2003, 325). Few championed Darwin's theory that adaptation occurs via differential reproduction.

In the second quarter of the twentieth century, a number of biologists argued that Darwin's theory of natural selection could unite research on adaptation with research on genetics. Early attempts at this synthesis include Fisher's *Genetical Theory of Natural Selection* (1930) and Haldane's *The Causes of Evolution* (1932). Later entries include Mayr's *Systematics and the Origin of Species* (1942) and Simpson's *The Meaning of Evolution* (1949).

While these biologists made their case to fellow biologists in rigorous academic publications, Julian Huxley brought the case for Darwin to the general reading public. *The Science of Life* (1930), coauthored with H. G. and G. P. Wells, and *Evolution: The Modern Synthesis* (1942) laid out the evidence for Darwin's theory clearly and accessibly, arguing that it is supported by ongoing research in genetics. The case for Darwin in the lab and on the street was won, and in the second half of the twentieth century, Darwin's name became synonymous with modern evolutionary biology (Gould 2002, 569–570).

The victory of the Modern Synthesis spelled the defeat of Lamarckism. The picture of adaptation that emerged is that only genes are heritable. So-called soft inheritance, the inheritance of acquired traits, was ruled out entirely (Gould 2002, 507).[2] A giraffe can stretch its neck all it wants, but it has no way of passing a stretched neck down to its offspring. The child inherits only the genes its parents were also born with. Adaptation occurs because creatures with "bad" genes are less likely to reproduce and multiply these genes, while creatures with "good" genes are more likely to reproduce and multiply these genes. What makes genes "good" or "bad" is the creature's ever-changing environment. If temperatures rise, the gene for thick fur can become a liability.

Darwin, however, believed in multiple mechanisms of adaptation. He believed that adaptation occurs via differential reproduction (natural selection) but also via the inheritance of acquired traits (Lamarckism).[3] In both *Origin* and *Descent*, natural selection features more prominently than Lamarckism. It is easy to read these works and forget Darwin's soft

[2] However, several recent studies suggest that acquired traits – and more specifically, acquired emotional responses – can be inherited. One study conditioned mice to fear a particular odor and found that their offspring feared this odor without conditioning (Dias & Ressler 2014). The authors proposed that chemical changes to sperm cells, attributable to the acquisition of this fear response, in turn modified the expression of genes in the offspring. A review of similar studies, evidencing the inheritance of acquired traits through epigenetic mechanisms, was titled "Lamarck Rises from His Grave" (Wang et al. 2017). This research remains controversial.

[3] I discussed Darwin's acceptance of the inheritance of acquired traits in §1.5.1.

spot for soft inheritance. In *Expression*, however, the inheritance of acquired traits is front and center, while differential reproduction is mentioned only in passing. It is difficult to read this work and forget Darwin's Lamarckism. (This, you may recall, is the first puzzle of *Expression* from §1.5.1: Why did Darwin rely on heritable habits instead of natural selection?)

Several scholars have suggested that Darwin's commitment to soft inheritance in *Expression* can be chalked up to a stubborn memory.[4] Darwin came up with his principles of expression in 1838, prior to discovering natural selection, when Lamarckism was still the best theory in town.[5] Once Darwin developed his theory of natural selection, he should have updated those principles of expression. Regrettably, he didn't. He included them in *Expression*, thereby smuggling Lamarckism into one of his most popular and interesting books.

What this reconstruction obscures, I contend, is that Darwin remained committed to heritable habits, despite his discovery of natural selection, because heritable habits were the linchpin of his evolutionary associationism.[6] Associationism posited habit as the basis of all thought. Associationism was still, in the late 1860s and early 1870s, the dominant theory of emotion. Had Darwin "updated" his principles of expression to remove all mention of heritable habits, then *Expression* would have been received as bad psychology. Critics could have charged that Darwin had failed to engage with recent advances in the philosophy and science of emotion.[7] The inclusion of heritable habits enabled Darwin to present *Expression* as a cutting-edge evolutionary contribution to the reigning theory of emotion in psychology, associationism.

In sum, because the Modern Synthesis rejected heritable habits, biologists were keen to forget that Darwin had embraced them. It was easier to view *Expression* as a puzzling anachronism than to remember that heritable habits were central to Darwin's attempt to unify biology and psychology. I turn now from the fall of Lamarckism in biology to the fall of associationism in psychology.

4 Montgomery (1985, 38) offers this explanation. Browne (2002, 69) and Barrett (2015, 67) suggest it.

5 I documented the genesis of Darwin's theory in Chapter 3, especially §3.2.

6 Radick (2010, 184–185) agrees that Darwin's acceptance of soft inheritance was more than a stubborn memory, but his explanation does not include Darwin's associationism.

7 Herbert Spencer, who independently developed a remarkably similar theory in the 1850s, presented his theory as a synthesis of associationism and physiognomy. I presented Spencer's theory in §§3.8.1 and 4.3.

6.3 Functional Psychology

When Darwin published *Expression* in 1872, associationism was still at the forefront of psychology. By the end of the century, however, it was largely rejected by leading psychologists. The decline and fall of associationism merits many pages of exploration, but I will focus on just two characters in this complicated story.[8] William James objected to associationism and developed a new approach to psychology, sometimes called functionalism. John Dewey claimed that Darwin was a forerunner of James's new school, rather than an adherent to the old school. This narrative stuck, and even today, Darwin's and James's theories of emotion are often mentioned in the same breath, as if kindred spirits. Darwin's contributions to associationism have been forgotten.

In his *Principles of Psychology* (1890), James rejects the foundational premise of associationism, namely, the idea that "our mental states are composite in structure, made up of smaller states conjoined" (1890, 148). Recall Alexander Bain's theory of emotion, which I presented in §3.8.2:

> the theory of Emotion will consist in the statement of the general laws that regulate the concurrence, coalition, transfer, or aggregation of the primary feelings – the Muscular states and the Sensations; and the analysis of the several Emotions will be the account of the component feelings in each, and the mode of their composition under some of these general laws. (Bain 1865, 35–36)

In other words, emotions are aggregates or compounds of simple sensations and actions, combined according to the principles of association. James rejects this entire metaphysics.

James's arguments against associationism are, quite frankly, abstruse and elusive.[9] His alternative to associationism is easier to articulate. If associationism is a bottom-up theory that starts with elements and works its way up to compounds, then James's functionalism is a top-down theory that begins with the rich fullness of experience and seeks to explain our ability to focus attention on particular aspects of experience. James writes:

[8] Commins (1932) details a turning of tides against associationism at the turn of the century. See also Young (1970).

[9] See Madden & Madden (1978) and Dunham (2023) for reconstructions of James's arguments. See Henle (1990), Woody (1999), and Dacey (2020, §2.c) for discussions of James's complex relationship to associationism. These scholars argue that despite James's harsh criticisms of associationism, his own theory preserves many features of associationism.

> Most books [on psychology] start with sensations, as the simplest mental facts, and proceed synthetically, constructing each higher stage from those below it. But this is abandoning the empirical method of investigation. No one ever had a simple sensation by itself. Consciousness, from our natal day, is of a teeming multiplicity of objects and relations, and what we call simple sensations are results of discriminative attention, pushed often to a very high degree. (James 1890, 224)

James proposes that we replace the old associationist language of "trains of thought" with a new functionalist language of "streams of thought." The former metaphor implies the conjunction of discrete entities; the latter metaphor implies a fluid flux of an amorphous substance (James 1890, 239). Indeed, James argues that thought is "in constant change" (1890, 229) and "sensibly continuous" (1890, 237), by which he means that we cannot normally identify the point at which one thought or feeling starts and the next begins. "[Consciousness] is nothing jointed; it flows" (1890, 239).

James does not discount association completely. He denies that association is the glue that fuses elementary thoughts and feelings together, but he insists that association plays an important role within our flowing thoughts. "*Association . . . is between* THINGS THOUGHT OF – *it is* THINGS, *not ideas, which are associated in the mind.* We ought to talk of the association of objects, not of the association of ideas" (1890, 554). The thought of salt leads naturally to the thought of pepper, not as two discrete mental states, but as a flowing stream of thought that shifts from the consideration of one object to a consideration of another, associated object.

At the end of his chapter on association, James sketches a brief history of associationism (1890, 594–603) that resembles mine from Chapter 2. Without focusing on the analysis of emotion, as I have, James surveys the theories of Hobbes, Hume, Hartley, Mill, Spencer, and Bain. James does not list Darwin as an associationist, nor does he acknowledge Darwin's contributions to associationism.[10] This point is important, as James and later Dewey attempt to portray Darwin instead as a forerunner of James's theory. We see the first traces of this strategy in James's theory of emotion.

[10] Titchener repeats James's history in a delightful footnote in his *Text-Book of Psychology*: "Let the roll be called! Thomas Hobbes, John Locke (who introduced the phrase 'association of ideas'), George Berkeley, David Hume, David Hartley (the founder of modern associationism), Thomas Brown, James Mill (the typical representative of the school), John Stuart Mill, Alexander Bain, Herbert Spencer (these three no longer pure associationists)" (1928, 374–375).

In what is arguably the most famous essay in all of emotions research, "What Is an Emotion?" (1884), James presents a counterintuitive theory of emotion:

> Our natural way of thinking about these standard emotions is that the mental perception of some fact excites the mental affection called the emotion, and that this latter state of mind gives rise to the bodily expression. My thesis on the contrary is that *the bodily changes follow directly the PERCEPTION of the exciting fact, and that our feeling of the same changes as they occur IS the emotion.* Common sense says, we lose our fortune, are sorry and weep; we meet a bear, are frightened and run; we are insulted by a rival, are angry and strike. The hypothesis here to be defended says that this order of sequence is incorrect, that the one mental state is not immediately induced by the other, that the bodily manifestations must first be interposed between, and that the more rational statement is that we feel sorry because we cry, angry because we strike, afraid because we tremble. (1884, 189–190)

Common sense says that the perception of an exciting fact causes a feeling, which in turn causes bodily changes. James's theory is that the perception of an exciting fact causes bodily changes, which in turn cause a feeling. Thus, James flips the causal order of feelings and bodily changes.[11]

James must now explain how the perception of an exciting fact can directly cause bodily changes. In the commonsense view, emotional feelings are inherently motivational, and so it makes sense that they would cause bodily changes. But the idea that we see a bear and then start to run, without having to fire up the fear engine, calls for an explanation. At this point, James appeals to evolution and adaptation: "readers of this Journal do not need to be reminded that the nervous system of every living thing is but a bundle of predispositions to react in particular ways upon the contact of particular features of the environment" (1884, 190). Citing Darwin, he describes every living creature as a lock, and aspects of its environment as keys (James 1884, 191). When a dog sees an enemy, it immediately tenses its muscles in preparation to fight. When a dog sees its master instead, it immediately relaxes those same muscles. In every case, the bodily changes that we normally refer to as "emotional expressions" can instead be explained as instinctual responses to environmental stimuli. They follow directly upon the perception of the exciting fact.

We find here the first draft of a story that Dewey will later refine. Darwin's theory removes emotions from the explanation of behavior,

[11] For interpretations of James's theory, see Dewey (1894), Cannon (1927), Ellsworth (1994), Prinz (2004), and Pott (2024).

asserting that so-called expressions are, rather, instinctual responses to environmental stimuli. (This, you may recall, is the second puzzle of *Expression* from §1.5.2: Why does Darwin sidestep emotion in the explanation of expression?) James's theory finds a new place for emotions in this explanation, asserting that emotions are simply the feelings of these instinctual responses as they occur. In sum, James complements Darwin's theory of emotional *expression* with his own theory of emotional *experience.*

Ten years after James published "What Is an Emotion?" (1884), John Dewey published two essays on James's theory, entitled "The Theory of Emotion (1)" (1894) and "The Theory of Emotion (2)" (1895). Dewey begins the first essay by noting that despite James's appeal to Darwin, there is a prima facie tension between their theories:

> The necessity of bringing the two theories together may be seen from the fact that the very phrase "expression of emotion," as well as Darwin's method of stating the matter, begs the question of the relation of emotion to organic peripheral action, in that it assumes the former as prior and the latter as secondary. Now this assumption, upon the basis of [James's theory], is false. (Dewey 1894, 553)

In other words, Darwin's theory appears to assume the commonsense view that emotional feelings precede emotional expressions.[12] Recall Darwin's first principle, the principle of serviceable associated habits. If an animal develops a habit of bristling whenever it has a particular feeling, then this feeling will cause it to bristle, regardless of whether bristling serves any purpose in that context. Darwin's second principle, the principle of antithesis, likewise seems to imply that feelings come before expressions. If an animal tenses up when afraid, then upon feeling the opposite (relief), it will involuntarily do the opposite (loosen up). First the feeling, then the expression. In his discussion of the facial feedback hypothesis, Darwin allows for the sequence to be reversed, with the voluntary performance of a bodily change intensifying the feeling.[13] Normally, however, the feeling comes first. Yet this is exactly what James's theory denies. According to James, an emotion just is the feeling of what we call the "expression," and so the feeling *cannot* occur before the expression. The worry, then, is that it may be impossible to reconcile James's and Darwin's theories. We cannot accept one without rejecting the other.

Appearances notwithstanding, Dewey insists that there is a way of reconciling these two theories. He argues that if we look past Darwin's

[12] Mead (1967, 17), for instance, attributes this view to Darwin.
[13] I discussed this hypothesis in §§1.3 and 1.6.

casual, uncritical use of the phrase "expression of emotion," we can see that Darwin's theory in fact *requires* bodily changes to spring directly from the perception of the exciting fact, without the interpolation of a feeling. He asks his reader to consider "whether in every case the idea of the expression of emotion does not enter in only to confuse" (1894, 555). Dewey concludes not merely that the two theories are compatible, but that Darwin's theory supports James's theory.

Dewey makes his case by considering each of Darwin's principles. Of the first, the principle of serviceable associated habits, Dewey points out that "the principle of explanation *actually* used, whatever the form of words employed, is that of survival, in the form of attitudes, of acts originally useful not *qua* expressing emotion, but *qua* acts – as serving life" (1894, 555). In other words, Darwin's first principle shows that some attitudes originated as useful responses to stimuli, and continue to be responses to those stimuli, even if no longer or not always useful. An animal perceives an enemy. It voluntarily bristles to make itself appear larger. It does this repeatedly, until a habit forms. Now, whenever this animal perceives an enemy, it bristles involuntarily. Perhaps this habit is inherited as an instinct by its offspring. Notice that I have not included any reference to feeling here. As a matter of fact, Darwin does insert feeling into the equation. He writes that the behavior is an expression of the feeling. Dewey's point, however, is that feelings don't add anything to Darwin's explanation. If an animal perceives an exciting fact, bristles, and forms a habit, then it will bristle whenever it perceives this or a similar exciting fact. Feelings may also occur, but there is no reason for them to come before the bristling. Feelings could come after the bristling. Dewey concludes that "*The reference to emotion in explaining the attitude is wholly irrelevant*" (1894, 556).

Dewey pursues the same line of argumentation regarding the second principle, the principle of antithesis. Darwin suggests that, once a habit has formed, an animal will instinctually perform an opposing movement upon experiencing an opposing feeling. The appeal to feeling is, however, redundant. Darwin could have said that once a habit has formed, an animal will instinctually perform an opposing movement upon *perceiving an opposing fact*. Wild animals, according to Dewey, have two fundamental attitudes: those related to finding food and those related to defending themselves against attackers (1894, 567). Domesticated animals have a third, "that of reception – the attitude of complete adaptation to something outside of itself" (1894, 567). Dewey suggests that there is a natural opposition between reception and defending. The sight of a foe prompts a defensive posture. The sight of a friend prompts a receptive posture. There is no reason to interpose an emotion between the perception and the attitude. Doing so would render the feeling redundant in the explanation.

Dewey extends his argument to the third principle as well, but we need not review those details here. The overarching point is becoming clear. Dewey's essay begins with the worry that James's and Darwin's theories may be incompatible. It proceeds with a demonstration that they are, in fact, compatible. And it ends with the suggestion that Darwin's theory all but entails James's. Dewey argues that while Darwin speaks the language of the old theory, in which perceptions lead to feelings and then actions, his spirit is with the new theory, in which feelings come last.

James and Dewey succeeded in establishing this narrative of Darwin as a proto-functionalist. Open any psychology textbook to the chapter on emotion, if it has one, and you are likely to find a version of the following story.[14] Charles Darwin kicked off the psychology of emotion by identifying universal behaviors characteristic of different emotions. He did not, however, articulate a theory of emotion per se. William James filled in the gap by suggesting that emotions are simply the feelings of those behaviors that Darwin identified. This narrative casts Darwin as a precursor to James and presents James as one of the first modern psychologists. Darwin's contributions to an older school of psychology are absent.

I have so far traced the forgetting of Darwin's philosophy of emotion across changes in biology and psychology. But changes in philosophy, too, are to blame.

6.4 Analytic Philosophy

Sometime around the turn of the twentieth century, a new philosophical tradition emerged, spreading quickly through the English-speaking world.[15] Analytic philosophy, as it would be called, proposed a clear division between empirical (a posteriori) and conceptual (a priori) truths. Philosophers would defer to scientists on empirical questions while claiming jurisdiction on conceptual questions. Conceptual analysis had always been *a* method of philosophy, but in this tradition, it became *the* method of philosophy. Analytic philosophers approached the philosophy of emotion very differently from the philosophers who had come before them.

[14] See, for example, Workman & Reader (2014, 331).
[15] See Soames (2003a; 2003b) and Schwartz (2012) for recent reconstructions of this history.

William James's philosophy of emotion included both empirical and conceptual claims. James hypothesized that bodily expressions come before emotional feelings (an empirical claim), and he insisted that an emotion *just is* the feeling of bodily expressions (a conceptual claim). The associationist philosophers likewise affirmed both empirical and conceptual claims about emotion. They hypothesized that perceptions can set in motion trains of associated thoughts, feelings, and actions (an empirical claim), and they insisted that an emotion *just is* the whole unfolding sequence (a conceptual claim).

Analytic philosophers of emotion proposed a strict division of labor. Psychologists are in charge of answering the empirical questions concerning emotion. Philosophers are in charge of answering the conceptual questions. The philosopher of emotion Robert Solomon put the point as follows in a volume showcasing different disciplinary approaches to emotion:

> What is an emotion? Because philosophy is a discipline concerned with the essential nature and the "definition" of things, the basic question facing theories of emotion in philosophy is still the question posed by James and answered, in a fashion, by Aristotle. It is, on the face of it, a quest for a definition, a conceptual analysis.... Which of the evident aspects of emotion – that is, the various sensory, physiological, behavioral, cognitive, and social phenomena that typically correspond with an emotion – should we take to be essential? (Solomon 2008, 10)

Granted, emotions normally involve thoughts, feelings, and actions, but which of these elements is essential to emotion (Prinz 2004, 18)? *That* is the philosophical question.

By the middle of the twentieth century, analytic philosophers of emotion had formed a consensus: Emotions are essentially thoughts, or cognitions (Deigh 1994). Yes, emotions tend to feel a particular way and tend to make us act in particular ways, but neither of these components is essential to the emotion. It does not even matter how these different components are causally connected. Regardless of whether the thought occurs first or last, or somewhere in between, the thought is the essential ingredient.

The reasoning behind this view runs as follows. Emotions are intentional states, or states directed toward and about objects and states of affairs (Kenny 1963; Pitcher 1965). We say that a person is afraid *of the bear*, or sad *that the stock market crashed*, or happy *about their promotion*. The object can be the emotion's cause, but it doesn't have to be. After all, we can fear monsters that don't exist and grieve losses that we only imagine.

Emotions do not simply have objects; they represent those objects in particular emotion-specific ways (Kenny 1963, 132–134). A person is afraid *of the bear* because they represent the bear *as dangerous*. They are sad *that the stock market crashed* because they represent this crash *as a loss*. And they are happy *about their promotion* because they represent the promotion *as a win*. More generally, we can say that whatever a person is afraid of, that person must represent it as dangerous. Whatever they are saddened by, they must represent it as a loss. And whatever they are happy about, they must represent it as a win.

Emotions represent objects in particular ways. This fact makes emotions very different from simple feelings (Pitcher 1965). When I feel an itch, or an ache, or a pain, I am not representing an object or state of affairs in a particular way. Instead, I seem to be reacting to a stimulus. Because emotions represent objects in particular ways, they are more like beliefs. I believe *of the Earth* that it *is round*. Emotions, like beliefs, follow a logic of predication (Solomon 1977).

It is because emotions represent objects in particular ways that we often evaluate them as rational or irrational (D'Arms & Jacobson 2000; Goldie 2000, 34). It's rational to fear a bear because in representing the bear as dangerous, we represent it truly. Bears are indeed dangerous. It's irrational to fear a rabbit because in representing the rabbit as dangerous, we represent it falsely. Rabbits are not dangerous.

Notice that this analysis draws bright lines between emotion types. The difference between fear and sadness is as simple and straightforward as the difference between danger and loss. Every emotion type corresponds to a unique way in which a subject can represent an object. Emotion types may be further divided into emotion subtypes. If anger represents something as an *offense*, then indignation may represent something as an *offense against others*, while resentment may represent something as an *offense against oneself* (Strawson 1962).

Notice also that this analysis presumes that it's the job of philosophers, not psychologists, to demarcate emotion types. Psychologists can document the feelings, physiological changes, and expressions that accompany emotions. They can determine the order of events in the causal chain. But empirical methods cannot determine the border of sadness and fear. You need an analytic philosopher for that.

Once philosophers had reduced the philosophy of emotion to the conceptual analysis of emotion types, it became increasingly difficult to view associationist theories of emotion as *philosophical* theories. These theories posited a causal mechanism for linking thoughts, feelings, and behaviors together. They called the whole train of associations the emotion, and they refused to identify any particular component as the "essential" component of the emotion. Because the associationists were not playing the game, as defined by twentieth-century analytic philosophers of emotion, they were easily forgotten. Histories of the philosophy of emotion either disregard the empiricist tradition or mislabel the empiricists as simple "feeling" or "noncognitive" theorists who identify emotions with simple, unanalyzable feelings (e.g., Scarantino & De Sousa 2018, §3).

Once the associationist theories of emotion were forgotten, Darwin's contributions to these theories were likewise forgotten.

I have argued so far that changes across three disciplines – biology, psychology, and philosophy – explain why Darwin's innovative philosophy of emotion has been neglected. *Expression* came to be read as a purely "scientific" exploration of emotional expression, and an outdated one at that, which retreated from philosophical questions concerning the emotions themselves. No one remembered that Darwin wrote *Expression* as an informed contribution to a rich philosophical literature on the nature of mind and emotion.

6.5 Conclusion

My aim in this book has been to resurrect Darwin's forgotten philosophy of emotion. In the course of solving three puzzles of *Expression*, I have defended five theses:

1. Darwin's *Expression* is a work of associationist philosophy.
2. Darwin analyzes emotions as trains of associations.
3. Darwin categorizes expressions as parts of emotions.
4. Darwin denies that emotions are adaptations.
5. Darwin claims that emotion categories are merely conventional.

These five theses highlight the relevance of Darwin's theory to contemporary emotions research.

First, **Darwin's *Expression* is a work of associationist philosophy**. Histories of philosophy generally tell one story about British Empiricism. This story centers on an epistemological theory – that all knowledge derives from experience – and traces the development of this theory through Hobbes, Locke, Berkeley, and Hume.[16] But there is another interesting story to tell. This story centers on a theory of emotion – that

[16] See, for instance, Bertrand Russell's *History of Western Philosophy* (1946), Charles Leander Hill's *A Short History of Modern Philosophy from the Renaissance to Hegel* (1951), and Frederick Copleston's *A History of Philosophy* (1958).

emotions derive from experience – and traces the development of this theory beyond Hume, through the associationists: Hartley, Bain, Spencer, and both Darwins. By the end of this story, empiricist philosophy and empirical psychology have become intertwined.

My story of British Empiricism also begins with Hobbes and Locke, who argue that all our ideas, including our emotions, originate in experience. They propose that emotions are "trains of ideas," prompted by sensations. Hobbes analyzes anger, for example, as a rapid train of ideas that includes the thought of being hurt by another, an aversion to that injury, and the thought of resisting (2020, VI.14–18, p. 44). Locke analyzes anger as a feeling of uneasiness that follows the thought of receiving injury and the intent to seek revenge (1894, II.20.10, p. 306). These analyses serve the broader empiricist project by showing how emotions derive from experience. Anger is the lesson learned from "fool me once, shame on you; fool me twice, shame on me."

Hobbes and Locke only gesture toward the force that connects trains of ideas. Furthermore, their analyses of emotions are cursory. Hume adds clarity and detail by distinguishing among three principles of association – resemblance, contiguity, and cause and effect (2007, 1.1.4.2, p. 13) – and by charting the complex associative relations internal to emotions like pride and humility (2007, 2.1.2, pp. 182–184). We feel pride, for example, when we perceive something pleasurable that we associate via cause and effect with ourselves. The pleasure felt toward the object is then associated via resemblance to the pride felt about ourselves (Schmitter 2021b, §5).

Hartley makes an early attempt to integrate empiricist philosophy with empirical science. He proposes that philosophical theories of association ought to be grounded in empirical theories of the nervous system. His attempt is admittedly speculative, positing a simple correspondence between Lockean mental associations and Newtonian nerve vibrations (1749, 33–34). In a bout of anger, for example, thoughts of injury shift to thoughts of resistance when one nerve's vibrations cause an adjacent nerve to vibrate. Hartley's vision of unifying philosophy and science sets a new course for the empiricist project.

Erasmus Darwin and Alexander Bain write influential textbooks that promote Hartley's vision. Erasmus's *Zoonomia* (1796) adopts the associationist theory that emotions are trains of thought, feeling, and action, linked by associations, and then documents in great detail the specific physiological changes activated in episodes of fear, anger, guilt, and so on. He proposes two principles of association – corresponding to Charles

Darwin's principles of serviceable associated habits and antithesis – to explain how these physiological changes become linked to trains of thoughts and feelings. We tremble when afraid, for instance, because the traumatic moment of birth fixed in our minds a strong association between this sensation and motion (E. Darwin 1796, 16.8.1). Similarly, Bain's *The Emotions and the Will* (1859) includes lengthy quotations from anatomical texts detailing the specific muscles activated in emotional episodes, and then proposes additional principles of association – including relativity, harmony, and conflict – to link these muscle contractions to the thoughts and feelings characteristic of each emotion (Bain 1865, 35–36).

The philosophers discussed thus far are pre-evolutionary empiricists. Herbert Spencer and Charles Darwin are the first evolutionary empiricists. Darwin writes: "Plato . . . says in Phaedo that our 'necessary ideas' arise from the preexistence of the soul, are not derivable from experience – read monkeys for preexistence" (2009, M 128). In other words, all ideas derive from experience, but not all ideas derive from our own personal experiences. Some ideas derive from our ancestors' experiences. Spencer (1855) and Darwin (1872) both analyze emotions as trains of thought and action guided by *inherited* associations. To explain why we turn red in anger or white in fear, we must examine the lives of our distant animal ancestors. As Darwin puts it: "With mankind some expressions . . . can hardly be understood, except on the belief that man once existed in a much lower and animal-like condition" (1872, 12). The turn to evolutionary empiricism is philosophically profound. However, for reasons I documented at the beginning of this chapter, Spencer's and Darwin's project is quickly abandoned. Emotions research splinters in different directions.

My story of British Empiricism enriches our present understanding of the history of emotions research. Against the familiar narrative that Darwin merely documented empirical facts about expression without attending to philosophical assumptions about emotion, and that William James built the first modern scientific theory of emotion on those facts, we see that Darwin's *Expression* was a contribution to a rich philosophical tradition attempting to integrate empiricist philosophy with empirical psychology. Anthologies of modern philosophy could include excerpts from Darwin's notebooks and *Expression* alongside other empiricist classics. Textbooks of psychology could identify Hobbes, Hume, and Hartley as antecedents to Darwin's evolutionary theory of expression.

The second thesis that I have defended in this book is that **Darwin analyzes emotions as trains of associations**. Emotions are not singular mental states; they are unfolding sequences of feelings, thoughts, and actions. Fear is the perception of a bear, followed by the thought that this bear as dangerous, followed by a painful feeling of excitement, followed by a bloodcurdling scream. Fear is not one element; it is the whole sequence. This analysis of emotion is not unique to Darwin, of course. He inherited it from his empiricist forebears and added an evolutionary twist.

Darwin's view should be of great interest to contemporary philosophers, since it challenges long-standing assumptions about the ontology of emotion. For much of the twentieth century, analytic philosophers of emotion analyzed emotions as mental *states* – though they disagreed as to whether emotions are essentially cognitive, perceptual, affective, somatic, motivational, or some other kind of state. Recently, however, some philosophers have suggested that emotions may instead be mental *events* (Robinson 2018; Naar 2024; Rowson forthcoming).[17] The difference between states and events is subtle but significant: They occupy time in two fundamentally different ways (Steward 1997, 73).

Events are *occurrents*, or things that "*extend* through time, that is, have different temporal *parts* at different times" (Naar 2024, §2). Think of an event like a movie showing. A movie showing normally stretches out over a couple hours. The first act occurs first, then the second act, then the third. Finally, the credits roll. The movie is made up of many temporal parts, or parts that occupy distinct slices of time. You haven't really seen the movie until you've seen all the parts, ideally in their proper order.[18]

States, by contrast, are *continuants*, or things that "lack temporal parts and are thus 'wholly present' at every point of time during which they exist" (Naar 2024, §2). Think of a person. Although I change over time, becoming older, grayer, and of course wiser, I am "wholly present" throughout the aging process. In this way, a person occupies time differently from a movie. Persons "are 'in full view' at any given time, with no parts missing" (Naar 2024, §2). Movies are "in partial view" at any given time, with their other temporal parts having already or not yet occurred. States occupy time in the same way that persons do. For as long as I am in

[17] Robinson (2018, 51) in fact defines emotions as "processes," but following Naar (2024) I use the term "events" to designate what Robinson means by "processes."

[18] Events and processes are closely related. According to one promising account, "events are temporal particulars and [processes] are temporal stuffs" (Rowson forthcoming, §4.1). In other words, processes are related to events in the way that gold is related to gold nuggets and gold jewelry. Processes are the "stuff" out of which events are made. For example, the event *John watched TV yesterday* is made up of the process *John watching TV*. The process fills time, and the event is what we get when we delineate a particular chunk of time, stretching from when John started watching TV to when he stopped watching TV.

the state of *believing that p*, this belief is "wholly present" and "in full view." No temporal parts are missing.[19]

Events have temporal parts. States do not. However, states can be temporal parts of events. The event *John forgot that p* began with John being in the state of *knowing that p* and ended with John being in the state of *no-longer-knowing that p*. The event unfolds over time; the states are fully present for as long as they obtain.

Darwin's theory is that emotions are mental *events*, which nonetheless have mental *states* as temporal parts. This theory opens up new and interesting questions about the nature of emotion. For example, are emotions "simple" or "complex" events (Rowson forthcoming, §4.1)? Complex events aim toward a resolution and generally pass through a series of stages to reach it. Simple events are aimless and generally stageless. A *basketball game* is a complex event because it progresses through multiple periods or quarters until a team accrues more points and is declared the winner. A *dance party* is a simple event because there is no goal beyond dancing and the party ends only when the dancers lose interest or pass out.

Recently, several philosophers have suggested that emotions are complex events (Robinson 2018; Na'aman 2020; 2021; Rowson forthcoming, §4.2). The stages of grief aim toward acceptance. The stages of anger aim toward destruction or, in some cases, social justice.[20] Darwin likewise seems to think that many emotions aim toward, and find resolution in, their characteristic behavioral expressions (e.g., Darwin 2009, N 66).

That being said, Darwin's view of emotions as trains of associations points us to the possibility that some emotions may be simple events. In other words, some emotions may be trains of associations that have no natural resolution and can in principle fill out extended periods of time.

19 States and dispositions are closely related. There is a long and checkered history in the philosophy of mind and psychology of defining mental states in terms of dispositions. To be *thirsty* is to be disposed to drink water when water is available. To *know that p* is to be disposed to answer "p" when asked. To *be afraid of bears* is to be disposed to scream and run away from bears. The precise difference between mental states and dispositions remains a matter of thorny debate, but it seems generally true that being in a mental state disposes one to think, feel, or do various things. This vague claim is sufficient for our purposes. See Fodor (1981) for a brief and accessible overview. See Kind (2020) for a more detailed overview. Ryle's *The Concept of Mind* (2009) is the *locus classicus* for logical behaviorism, or the attempt to reduce mental states to behavioral dispositions. Ryle's chapter 4 analyzes emotion. McGill & Welch (1946) develop an early behaviorist analysis of emotion.

20 Philosophers continue to debate the merits of anger. Some are critical on account of anger's propensity to incite violence (Pettigrove 2012; Nussbaum 2016). Others are supportive on account of anger's capacity to motivate agents to pursue social change (Lorde 1984, 124–133; Bell 2009; Srinivasan 2018; Cherry 2021).

Indeed, some emotions are characterized by *ruminative inertia* (Nolen-Hoeksema 2000; Koval et al. 2012; Bean et al. 2020; Bean et al. 2021). Ruminative inertia is the combination of two conditions: *rumination*, or the repetition of similar or related thoughts, and *emotional inertia*, or the tendency of an emotion to persist into the next moment. Ruminative inertia is therefore the tendency of an emotion to persist in virtue of the repetition of similar or related thoughts.[21] Worrying is a prime example. It is a process that can fill two minutes, two hours, or even two days. Worrying does not seem to aim toward any particular resolution. It is more like a dance party than a basketball game. Darwin's view invites us to consider that different emotions may fill out time differently and have different temporal shapes.

The third thesis that I have defended is that **Darwin categorizes expressions as parts of emotions**. Mental states are temporal parts of emotions, but physical reactions are too. In the event of fear, we perceive something as dangerous, but we also shriek and run away. Darwin writes that "ideas *&* *bodily actions* make the emotions" (2009, N 89; emphasis added).

The view that expressions are parts of emotions features prominently in debates over our knowledge of other minds. Some philosophers claim that we must infer emotions from their expressions, while others claim that we can perceive emotions directly in their expressions.[22] The latter often defend this claim by asserting (1) that expressions are parts of emotions and (2) that seeing a part of something is generally sufficient for perceiving that something.[23] If I see my dog's tail sticking out from under the sofa, then I've seen *my dog*. I don't need to see the rest of him. Analogously, if I see the perceptible part of an emotion, namely, its expression, then I've seen *that emotion*. I don't need to see the other, imperceptible parts.

Recently, Rowson (forthcoming, §3) has identified a flaw in this line of reasoning. The problem is that emotions do not have parts in the same way that ordinary objects have parts. My dog's tail is a *spatial* part of him. It occupies the space directly behind his behind. His tail is normally a fair distance from his teeth, but when he chases his tail, these parts meet. Ordinary objects have spatial parts. Emotions do not. More to the point:

[21] Charles Darwin's notebooks contain reflections on Erasmus Darwin's theory that emotions are a mild form of insanity or delirium, since both are trains of thought that can occur against one's will. Charles discusses Erasmus's theory in an 1838 notebook (2009, M 13–21). He discusses a similar theory of insanity in a notebook entry from around this same time or shortly thereafter (2009, OUN 17).

[22] I discussed these views in §§5.3 and 5.4. Important contributions to the debate include Morton (1980), Gordon (1986), Goldman (1989), Carruthers (1996), and Gallagher & Zahavi (2012).

[23] See Tormey (1971, 47–48), Hampshire (1976, 74–75), Green (2007, 84–93; 2010a), and Glazer (2018a).

Expressions are not spatial parts of emotions. There is no actual distance separating a pout from a feeling of dejection or a thought of profound loss. There is no way for this distance to shrink or grow.

Since expressions are not spatial parts of emotions, the perception of emotion cannot be modeled on the perception of ordinary objects. Rowson (forthcoming, §3) argues that a more promising strategy is to model the perception of emotion on the perception of events. "Did you see the game last night?" "Yes, I caught the second half." "Did you watch the parade?" "Yes, I watched a little bit of it." Events are occurrents; they stretch out over time and have *temporal* parts. Two people might watch different parts of an event, yet are watching the same event.

To model the perception of emotion on the perception of events, however, we must abandon the view that emotions are singular mental *states* and embrace the view that emotions are unfolding mental *events*. States do not have temporal parts, but mental events do. Expressions can only be temporal parts of emotions if emotions are events. This argument reveals a surprising point of contact between old empiricist theories of emotion and recent phenomenological theories of social cognition: If one wishes to defend the view that expressions are parts of emotions, then one may have to adopt something like Darwin's theory of emotion.

The fourth thesis is that **Darwin denies that emotions are adaptations**. Modern Darwinian theories of emotion tend toward adaptationism. That is, they tend to assume that emotions evolved because they promote survival and reproduction.[24] Disgust protects us from pathogens. Amusement motivates us to find camaraderie. Anger improves our bargaining position. Emotions are not always beneficial, of course. If your anger makes you punch your boss in the nose, then your material, social, and psychological conditions will likely worsen.[25] But on the whole, we have emotions because it is good to have emotions.

Darwin did not share this view. To the contrary, he speculated that emotions are, by and large, purposeless products of association.[26] They are neither good nor bad; they are neutral. Once upon a time, an animal

[24] I defended this interpretation in §4.4. See Plutchik (1962), Ekman (1992), Griffiths (1997), Frank (1988), Prinz (2004), Van Kleef (2009), and Hutto et al. (2018).

[25] Elsewhere I have argued that emotion regulation is adaptive for this very reason (Glazer 2020; 2021). Along these lines, Marco Viola (2022) proposes that sunglasses sometimes function as a socio-affective artifact that obscures spontaneous facial expressions of emotion.

[26] It is generally accepted that Darwin viewed emotional *expressions* as functionless (Burkhardt 1985, 359; Fridlund 1992; Campbell 1997a, 454; Richards 2009, 114). In §4.4, I argued that Darwin's anti-adaptationist stance generalizes to whole *emotions*.

caught, killed, and consumed its prey. Associations formed. The next time it encountered prey, its mind raced through these associations. It anticipated the thrill of the chase, the pain of the fight, and the satisfaction of the meal. It extended its claws as if gripping its prey, and it gnashed its teeth as if biting into it. The animal replayed the events of the prior hunt as an appetizer to the present hunt. This train of associated sensations and expressions is an emotion, an emotion that, in this case, serves no purpose. Despite serving no purpose, the associations may be passed down, generation after generation, to this animal's many descendants, belonging to new and diverse species, many of whom derive no benefit from replaying the sensations and expressions from their ancestor's successful hunt.

This story illustrates Darwin's general view of emotion. Emotions are "trains of thought long associated with action" (DAR LIB 404, 333) – sequences of thoughts, feelings, and actions, initiated by sensations, guided by associations. That distinctive feeling of anger that we all know is felt not because it is useful to feel it but because that feeling happened to accompany the perception of an obstacle long, long ago, and became strongly associated with it. That recognizable snarl of anger is shown not because it is useful to show it but because this expression is strongly associated with a feeling. Darwin did not assume that each part of an emotional sequence has earned its place by providing a survival benefit. To the contrary, Darwin assumed that the force of habit is strong enough to preserve myriad quirks and tics, provided that they are not so disadvantageous as to be selected against. Darwin's creationist adversaries looked for utility in every trait as proof of an intelligent designer. Darwin was interested in emotions precisely because he saw them as exceptions to the rule.

Darwin surely took his anti-adaptationism too far. The modern view that many features of emotions evolved because they promote survival and reproduction is most likely true. Still, some modern Darwinians have taken their adaptationism too far in the opposite direction. The increasingly popular view in philosophy that all natural expressions evolved as social signals, designed to communicate emotions to observers, may be guilty of this trend (see §5.7). Darwin's theory encourages us to consider that some natural expressions may not be signals. They may provide us with information about emotions, but incidentally, by association rather than by design. Blushing expresses embarrassment regardless of whether it evolved for this very purpose (Scarantino 2019, 58). Expressions can be happy accidents of association and evolution. Other features of emotion may be happy accidents, too.

The final thesis that I have defended in this book is that **Darwin claims that emotion categories are merely conventional**. According to basic emotions theory, emotions like anger and fear are natural kinds. Angers naturally belong together because they are produced by the same underlying mechanism, which evolved for its reproduction and survival value. Fears naturally belong together because they are produced by a different underlying mechanism, which likewise evolved for its evolutionary benefits. Granted, these natural emotions do not correspond perfectly to folk concepts of "anger" and "fear," which encompass a broad range of phenomena (Griffiths 1997). When I say that "they are afraid of the stock market crashing," I may not be describing an output of one of these basic emotion mechanisms. Rather, I may be describing a complex pattern of thoughts, feelings, and behaviors that only loosely resembles the immediate panic that occurs when one comes face-to-face with danger. Within the familiar folk concept of "fear," however, is a natural grouping of fears that form a natural kind.

Although basic emotions theorists claim to be walking in the footsteps of Darwin (e.g., Ekman 2009), I have argued that Darwin rejects the view that emotions form natural groupings.[27] Instead, he embraces Thomas Brown's view that "In nature, there are no classes" (Brown 1798, 321). All categories of emotion are, in Darwin's view, purely conventional, instituted for the sake of convenience. We group emotions together so that we may communicate more easily and efficiently about many distinct episodes of emotions all at once.

Because Darwin does not think that any groupings are more natural or objective than any other, he eschews the project of identifying a short list of basic emotions. Instead, he speaks of emotions in the diverse and varied terms with which we are most familiar. Again, this echoes Thomas Brown, who analyzes emotions "in those complex forms in which they usually exist, and have received certain definite characteristic names that are familiar to you" (1824, II.52, p. 255). Darwin enumerates common expressions of disgust, contempt, disdain, scorn, and loathing (1872, 254), without attempting to reduce or assimilate these emotion types to an Ur-type, as a basic emotions theorist may be inclined to do. For Darwin, these names denote loosely defined conventional groupings of emotions, none of which is more scientifically valid than any other. Darwin's *Expression* catalogues the longest list of emotions of any work of the period, naming more than fifty emotions.

[27] I argued for these points in §4.4.

In nature, there are no classes of emotions. Darwin maintains that every emotion – regardless of whether we call it "anger," "fear," or something else – is produced through the same underlying process, that of association. Our emotional lives are diverse and varied simply because the inputs to the process of association, including the sum of one's past associations, learned and inherited, are diverse and varied. One sensation sets in motion a train of associations that we call "anger," while another sets in motion a train of associations that we call "fear." Darwin's claim that emotion categories are conventional is provocative, but his theory that emotional variety can be explained with reference not to emotion kinds but to a single process of association plus the varying conditions of life, is philosophically profound.

Darwin's view anticipates a modern alternative to basic emotions theory, namely, psychological constructionism. Constructionists likewise claim that emotions are the products of simpler, emotion-independent processes, and that emotion terms pick out the resulting patterns insofar as they resemble mental prototypes. The irony of my interpretation of Darwin as a proto-constructionist is that many constructionists view Darwin as the inspiration for basic emotions theory, who led emotions research astray with his unreflective reliance on a folk theory of emotion.[28]

Charles Darwin writes in a biography of his grandfather, Erasmus Darwin, that "he had the true spirit of a philosopher" (Krause 1879, 49). Charles explains that natural philosophers ought to have two things: a keen eye and a powerful imagination. To uncover nature's secrets, one must observe carefully and document fastidiously. However, this is not enough. One must then follow a long train of thought connecting observed effects to unobserved causes, and particular events to general laws. Only those with the greatest power of imagination can follow these winding trains to their destinations. A true philosopher, in other words, has mastered the two faculties of mind posited by the associationist philosophers: perception and association.

I have argued in this book that Charles Darwin, no less than Erasmus Darwin, had the true spirit of a philosopher. *Expression* showcases Charles's keen eye for detail and vivid imagination. He reasons carefully and creatively from observed facts about emotional expressions to an ambitious philosophical theory of emotion. For this reason, and for the other reasons I have presented in this book, Darwin's theory of emotion should be of greater interest to contemporary philosophers, psychologists, and other theorists of emotion.

[28] See, for instance, Russell & Fernández-Dols (1997), Gendron & Barrett (2009), and Barrett (2013).

References

Abercrombie, J. (1838). *Inquiries Concerning the Intellectual Powers and the Investigation of Truth*, London: John Murray, Albemarle Street.

Abizadeh, A. (2017). Hobbes on mind: Practical deliberation, reasoning, and language. *Journal of the History of Philosophy*, 55(1), 1–34.

Allen, R. C. (2001). David Hartley's new words for action: "Automatic" and "decomplex." *Enlightenment and Dissent*, 20, 1–22.

Allport, F. H. (1924). *Social Psychology*, Boston: Houghton Mifflin.

Armon-Jones, C. (1986). The thesis of constructionism. In R. Harré, ed., *The Social Construction of Emotions*, Oxford: Blackwell, pp. 32–56.

Averill, J. (1980a). A constructivist view of emotion. In R. Plutchik & H. Kellerman, eds., *Theories of Emotion, vol. 1: Emotion: Theory, Research and Experience*, New York: Academic Press, pp. 305–339.

(1980b). Emotion and anxiety: Sociocultural, biological, and psychological determinants. In A. Rorty, ed., *Explaining Emotions*, Berkeley: University of California Press, pp. 37–72.

Bacon, F. (1901). *Novum Organum*, New York: P. F. Collier and Son.

Bain, A. (1855). *The Senses and the Intellect*, 1st ed., London: J. W. Parker and Son.

(1859). *The Emotions and the Will*, 1st ed., London: J. W. Parker and Son.

(1864). *The Senses and the Intellect*, 2nd ed., London: Longman, Green, Longman, Roberts and Green.

(1865). *The Emotions and the Will*, 2nd ed., London: Longmans, Green.

(1873). *Review of "Darwin on Expression": Being a Postscript to The Senses and the Intellect, London: Longmans, Green.*

Barber, M. E. (1867). Letter no. 5745. *Darwin Correspondence Project,* www.darwinproject.ac.uk/letter/?docId=letters/DCP-LETT-5745.xml (accessed September 28, 2023).

Bar-On, D. (2004). *Speaking My Mind*, New York: Oxford University Press.

(2010). Expressing as "showing what's within": On Mitchell Green's *Self-Expression* OUP 2007. *Philosophical Books*, 51(4), 212–227.

Barrett, L. F. (2006). Emotions as natural kinds? *Perspectives on Psychological Science*, 1, 28–58.

(2011). Was Darwin wrong about emotional expressions? *Current Directions in Psychological Science*, 20, 400–406.

(2012). Emotions are real. *Emotion*, 12, 413–429.
(2013). Psychological construction: A Darwinian approach to the science of emotion. *Emotion Review*, 5, 379–389.
(2015). Ten common misconceptions about the psychological construction of emotion. In L. F. Barrett & J. A. Russell, eds., *The Psychological Construction of Emotion*, New York: Guilford Press, pp. 45–79.
(2017a). *How Emotions Are Made: The Secret Life of the Brain*, New York: Mariner Books.
(2017b.) The theory of constructed emotion: An active inference account of interception and categorization. *Social Cognitive and Affective Neuroscience*, 12(1), 1–23.
Barrett, L. F., Adolphs, R., Marsella, S., Martinez, A. M., & Pollak, S. D. (2012). Emotional expressions reconsidered: Challenges to inferring emotion from human facial movements. *Psychological Science in the Public Interest*, 20(1), 1–68.
Barrett, L. F., & Russell, J. A. (2015). An introduction to psychological construction. In L. F. Barrett & J. A. Russell, eds., *The Psychological Construction of Emotion*, New York: Guilford Press, pp. 1–20.
Bean, C. A. L., Heggeness, L. F., & Ciesla, J. A. (2021). Ruminative inertia, emotion regulation, and depression: A daily-diary study. *Behavior Therapy*, 52(6), 1477–1488. https://doi.org/10.1016/j.beth.2021.04.004.
Bean, C. A. L., Heggeness, L. F., Kalmbach, D. A., & Ciesla, J. A. (2020). Ruminative inertia and its association with current severity and lifetime course of depression. *Clinical Psychological Science*, 8(6), 1007–1016.
Beatty, J. (1985). Speaking of species: Darwin's strategy. In David Kohn, ed., *The Darwinian Heritage*, Princeton, NJ: Princeton University Press, pp. 265–281.
Bell, C. (1801). *Engravings of the Arteries: Illustrating the Second Volume of the Anatomy of the Human Body, and Serving as an Introduction to the Surgery of the Arteries*, London: Printed by C. Whittington, Dean Street, Fetter Lane, for T. N. Longman and O. Rees, Paternoster-Row, and T. Cadell, Jun. and W. Davies, in the Strand.
(1802). *The Anatomy of the Brain, Explained in a Series of Engravings*, London: Printed by C. Whittington, Dean Street, Fetter Lane, for T. N. Longman and O. Rees, Paternoster-Row, and T. Cadell, Jun. and W. Davies, in the Strand.
(1803). *A Series of Engravings Explaining the Course of the Nerves*, London: Printed by C. Whittington, Dean Street, Fetter Lane, for T. N. Longman and O. Rees, Paternoster-Row; and T. Cadell, and W. Davies, Strand.
(1806). *Essays on the Anatomy of Expression in Painting*, 1st ed., London: Longman, Hurst, Rees, and Orme.
(1811). *Idea of a New Anatomy of the Brain – Submitted for the Observations of His Friends*, London: Strahan and Preston.
(1821). *Illustrations of the Great Operations of Surgery*, London: Printed for Longman, Hurst, Rees, Orme, and Brown, Paternoster-Row; by A. and R. Spottiswoode, Printers-Street.

(1824). *Essays on the Anatomy and Philosophy of Expression*, 2nd ed. London: John Murray, Albemarle-Street.

(1844). *The Anatomy and Philosophy of Expression as Connected with the Fine Arts*, London: John Murray.

Bell, M. (2009). Anger, virtue and oppression. In L. Tessman, ed., *Feminist Ethics and Social and Political Philosophy: Theorizing the Non-Ideal*, London: Springer, pp. 58–77.

Berkowitz, C. (2014). Charles Bell's seeing hand: Teaching anatomy to the senses in Britain, 1750–1840. *History of Science*, 42(4), 377–400.

(2015). *Charles Bell and the Anatomy of Reform*, Chicago: University of Chicago Press.

Bessey, C. E. (1908). The taxonomic aspect of the species question. *American Naturalist*, 42, 218–224.

Bird, A., & Tobin, E. (2023). Natural kinds. In E. N. Zalta, ed., *The Stanford Encyclopedia of Philosophy* (Spring 2023 edition), https://plato.stanford.edu/archives/spr2023/entries/natural-kinds/.

Bobier, C. (2020). Rethinking Thomas Hobbes on the passions. *Pacific Philosophical Quarterly*, 101(4), 582–602.

Bowler, P. J. (2003). *Evolution: The History of an Idea*, 3rd ed., Berkeley: University of California Press.

Boyd, R. (1999). Homeostasis, species, and higher taxa. In R. Wilson, ed., *Species: New Interdisciplinary Essays*, Cambridge, MA: MIT Press, pp. 141–186.

Briggs, J. L. (1970). *Never in Anger: Portrait of an Eskimo Family*, Cambridge, MA: Harvard University Press.

Broadie, A. (2003). The human mind and its powers. In A. Broadie, ed., *The Cambridge Companion to the Scottish Enlightenment*, New York: Cambridge University Press, pp. 60–78.

Brown, T. (1798). *Observations on the Zoonomia of Erasmus Darwin, MD*, Edinburgh: Mundell & Son.

(1805). *Observations on the Nature and Tendency of the Doctrine of Mr Hume Concerning the Relation of Cause and Effect*, Edinburgh: Mundell and Son.

(1820). *Lectures on the Philosophy of the Human Mind*, Edinburgh: James Ballantyne.

(1824). *Lectures on the Philosophy of the Human Mind*, 3 vols., Philadelphia: John Grigg.

Browne, J. (1985). Darwin and the expression of the emotions. In D. Kohn, ed., *The Darwinian Heritage*, Princeton, NJ: Princeton University Press, pp. 307–326.

(1995). *Charles Darwin: Voyaging*, Princeton, NJ: Princeton University Press.

(2002). *Charles Darwin: The Power of Place*, Princeton, NJ: Princeton University Press.

Buckingham, H. W., & Finger, S. (1997). David Hartley's psychological associationism and the legacy of Aristotle. *Journal of the History of the Neurosciences*, 6(1), 21–37.

Buffon (2007/1749–1767). *Œuvres*, S. Schmitt and C. Crémière, eds., Paris: Gallimard.

Burkhardt, R. W. (1977). *The Spirit of System: Lamarck and Evolutionary Biology*, Cambridge, MA: Harvard University Press.

(1985). Darwin on animal behavior and evolution. In D. Kohn, ed., *The Darwinian Heritage*, Princeton, NJ: Princeton University Press, pp. 327–366.

(2005). *Patterns of Behavior: Konrad Lorenz, Niko Tinbergen, and the Founding of Ethology*, Chicago: University of Chicago Press.

Cairns, C. (2007). *Associationism and the Literary Imagination: From the Phantasmal Chaos*, Edinburgh: Edinburgh University Press.

Campbell, S. (1997a). Emotion as an explanatory principle in early evolutionary theory. *Studies in History and Philosophy of Science*, 28(3), 453–473. https://doi.org/10.1016/S0039-3681(96)00018-0.

(1997b). *Interpreting the Personal*, Ithaca, NY: Cornell University Press.

Cannon, W. B. (1927). The James–Lange theory of emotions: A critical examination and an alternative theory. *American Journal of Psychology*, 39, 106–124. https://doi.org/10.2307/1415404.

Carruthers, P. (1996). *Language, Thoughts and Consciousness: An Essay in Philosophical Psychology*, New York: Cambridge University Press.

Chakravarrty, A. (2023). Last chance saloons for natural kind realism. *American Philosophical Quarterly*, 60(1), 63–81.

Charland, L. C. (2002). The natural kind status of emotion. *British Journal for the Philosophy of Science*, 53(4), 511–537. doi:10.1093/bjps/53.4.511.

Cherry, M. (2021). *The Case for Rage*, Oxford: Oxford University Press.

Clark, A. (2013). Whatever next? Predictive brains, situated agents, and the future of cognitive science. *Behavioral and Brain Sciences*, 36(3), 181–204.

Coan, J. A., & Gonzalez, M. Z. (2015). Emotions as emergent variables. In L. F. Barrett & J. A. Russell, eds., *The Psychological Construction of Emotion*, New York: Guilford Press, pp. 209–228.

Collier, M. (2011). Hume's science of emotions: Feeling theory without tears. *Hume Studies*, 37(1), 3–18.

Colp, R. (1986). Confessing a murder: Darwin's first revelations about transmutation. *Isis*, 77, 9–32.

Commins, W. D. (1932). Some early holistic psychologists. *Journal of Philosophy*, 29(8), 208–217. https://doi.org/10.2307/2016807.

Conrad, R. (1998). Darwin's baby and baby's Darwin: Mutual recognition in observational research. *Human Development*, 41, 47–64.

(2004). "[A]s if she defied the world in her joyousness": Rereading Darwin on emotion and emotional development. *Human Development*, 47, 40–65.

Copenhaver, R., & Odenbaugh, J. (2020). Experiencing emotions: Aesthetics, representationalism, and expression. In B. Brogaard & D. E. Gatzia, eds., *The Epistemology of Non-Visual Perception*, New York: Oxford University Press.

Copleston, F. J. (1959). *A History of Philosophy, vol. 5: British Philosophy*, London: Continuum.

Craig, W. (1921). A note on Darwin's work on the expression of the emotions in man and animals. *Journal of Abnormal Psychology and Social Psychology*, 16(5–6), 356–366.

Cranefield, P. F. (1974). *The Way In and the Way Out: François Magendie, Charles Bell, and the Roots of the Spinal Nerves*, Mount Kisco, NY: Futura Publishing.

Craver, C. F. (2009). Mechanisms and natural kinds. *Philosophical Psychology*, 22, 575–594.

Dacey, M. (2015). Associationism without associative links: Thomas Brown and the associationist project. *Studies in the History and Philosophy of Science Part A*, 54, 31–40. https://doi.org/10.1016/j.shpsa.2015.08.001.

(2020). Associationism in the philosophy of mind. In *The Internet Encyclopedia of Philosophy*, https://iep.utm.edu/associationism-in-philosophy-of-mind/.

(2022). The shared project, but divergent views, of the empiricist associationists. *Philosophical Psychology*, doi: 10.1080/09515089.2022.2060070.

D'Arms, J., and Jacobson, D. (2000). The moralistic fallacy: On the "appropriateness" of emotions. *Philosophy and Phenomenological Research*, 61(1), 65–90.

Darwin, C. (1858). Letter no. 2373. *Darwin Correspondence Project*, www.darwinproject.ac.uk/letter/?docId=letters/DCP-LETT-2373.xml (accessed September 28, 2023).

(1859). *On the Origin of Species by Means of Natural Selection, or Preservation of Favoured Races in the Struggle for Life*, London: John Murray.

(1863). Letter no. 4218. *Darwin Correspondence Project*, www.darwinproject.ac.uk/letter/?docId=letters/DCP-LETT-4218.xml (accessed August 23, 2022).

(1864). Letter no. 4378F. *Darwin Correspondence Project*, www.darwinproject.ac.uk/letter/?docId=letters/DCP-LETT-4378F.xml (accessed August 23, 2022).

(1865). Letter no. 4794. *Darwin Correspondence Project*, www.darwinproject.ac.uk/letter/?docId=letters/DCP-LETT-4794.xml (accessed August 23, 2022).

(1867a). *Queries on Expression*, London: n.p..

(1867b). Letter no. 5440. *Darwin Correspondence Project*, www.darwinproject.ac.uk/letter/?docId=letters/DCP-LETT-5440.xml (accessed October 4, 2023).

(1868a). *The Variation of Animals and Plants under Domestication*, London: John Murray.

(1868b). Queries about expression for anthropological inquiry. *Annual Report of the Board of Regents of the Smithsonian Institution*, Misc. Document No. 86 (1867), 324.

(1871). *The Descent of Man, and Selection in Relation to Sex*, London: John Murray.

(1872). *The Expression of the Emotions in Man and Animals*, London: John Murray.

(1873). Letter no. 9092. *Darwin Correspondence Project*, www.darwinproject.ac.uk/letter/?docId=letters/DCP-LETT-9092.xml (accessed August 23, 2022).

(1877). A biographical sketch of an infant. *Mind*, 2(7), 285–294.

(1958). *The Autobiography of Charles Darwin*, N. Barlow, ed., London: St. James Place.

(1980). *Metaphysics, Materialism, and the Evolution of Mind: The Early Writings of Charles Darwin*, H. E. Gruber & P. H. Barrett, eds., Chicago: University of Chicago Press.

(2009). *Charles Darwin's Notebooks, 1836–1844: Geology, Transmutation of Species, Metaphysical Enquiries*, P. H. Barrett, P. J. Gautrey, S. Herbert, D. Kohn, & S. Smith, eds., New York: Cambridge University Press.

Darwin, E. (1796). *Zoonomia; or, The Laws of Organic Life*, vol. 1, London: J. Johnson.

Daston, L., & Galison, P. (2007). *Objectivity*, Princeton, NJ: Princeton University Press.

Davidson, D. (2001). Hume's cognitive theory of pride. In D. Davidson, ed., *Essays on Actions and Events*, New York: Oxford University Press, pp. 277–290.

Davis, W. (1988). Expression of emotion. *American Philosophical Quarterly*, 25(4), 279–291.

(2008). Expressing, meaning, showing, and intending to indicate. *Intercultural Pragmatics*, 5(2), 111–129.

Deigh, J. (1994). Cognitivism in the theory of emotions. *Ethics*, 104(4), 824–854. doi:10.1086/293657.

(2004). Primitive emotions. In R. Solomon, ed., *Thinking about Feeling: Contemporary Philosophers on Emotions*, New York: Oxford University Press, pp. 9–27.

Demeter, T. (2021). Fodor's guide to the Humean mind. *Synthese*, 199, 5355–5375.

Desmond, A., Moore, J., & Browne, J. (2007). *Charles Darwin*, Very Interesting People Series, New York: Oxford University Press.

Dewey, J. (1894). The theory of emotion (I): Emotional attitudes. *Psychological Review*, 1, 553–569. https://psycnet.apa.org/doi/10.1037/h0069054.

(1895). The theory of emotion (II): The significance of emotions. *Psychological Review*, 2, 13–32.

Dias, B., & Ressler, K. (2014). Parental olfactory experience influences behavior and neural structure in subsequent generations. *Nature Neuroscience*, 17, 89–96. https://doi.org/10.1038/nn.3594.

Dickson, D. A., Paulus, J. K., Mensah, V., Lem, J., Saavedra-Rodriguez, L., Gentry, A., Pagidas, K., & Feig, L. A. (2018). Reduced levels of miRNAs 449 and 34 in sperm of mice and men exposed to early life stress. *Translational Psychiatry*, 8(1), 101. doi: 10.1038/s41398-018-0146-2.

Dixon, T. (2003). *From Passions to Emotions*, New York: Cambridge University Press.

(2012). "Emotion": The history of a keyword in crisis. *Emotion Review*, 4(4), 338–344. doi:10.1177/1754073912445814.

Duchenne, G. B. A. (1862). *Mécanisme de la physionomie humaine, ou analyse électro-physiologique de l'expression des passions. 1 vol. and "Atlas" of plates*, Paris: Ve Jules Renouard, Libraire.

Dunham, J. (2023). Flights in the resting places: James and Bergson on mental synthesis and the experience of time. *British Journal for the History of Philosophy*, 31(2), 183–204. https://doi.org/10.1080/09608788.2022.2136138.

Egerton, F. N. (1976). Darwin's early reading of Lamarck. *Isis*, 67, 452–456.

Ekman, P. (1980). Biological and cultural contributions to body and facial movement in the expression of emotions. In A. O. Rorty, ed., *Explaining Emotions*, Berkeley: University of California Press, pp. 37–72.

(1984). Expression and the nature of emotion. In K. Scherer & P. Ekman, eds., *Approaches to Emotion*, Hillsdale, NJ: Lawrence Erlbaum, pp. 319–343.

(1992). An argument for basic emotions. *Cognition & Emotion*, 6(3–4), 169–200.

(1994). All emotions are basic. In P. Ekman & R. Davidson, eds., *The Nature of Emotion: Fundamental Questions*, New York: Oxford University Press, pp. 15–19.

(2009). Darwin's contributions to our understanding of emotional expressions. *Philosophical Transactions B*, 364(1535), 3449–3451. doi:10.1098/rstb.2009.0189.

Ekman, P., & Friesen, W. V. (1978). *Facial Action Coding System (FACS)*. APA PsycTests. doi:10.1037/t27734-000.

Ekman, P., Sorenson, E. R., & Friesen, W. V. (1969). Pan-cultural elements in facial display of emotions. *Science*, 164, 86–88.

Elfenbein, H. A., & Ambady, N. (2002). On the universality and cultural specificity of emotion recognition: A meta-analysis. *Psychological Bulletin*, 128(2), 205–235.

Elfenbein, H. A., Beaupré, M., Lévesque, M., & Hess, U. (2007). Toward a dialect theory: Cultural differences in the expression and recognition of posed facial expressions. *Emotion*, 7(1), 131–146.

Ellis, B. (2001). *Scientific Essentialism*, Cambridge: Cambridge University Press.

Ellsworth, P. C. (1994). William James and emotion: Is a century of fame worth a century of misunderstanding? *Psychological Review*, 101(2), 222–229. doi:10.1037/0033-295x.101.2.222.

Fara, P. (2012). *Erasmus Darwin: Sex, Science, and Serendipity*, New York: Oxford University Press.

Fehr, B., & Sprecher, S. (2009). Prototype analysis of the concept of compassionate love. *Personal Relationships*, 16(3), 343–364.

Fernandez-Dols, J. M., & Ruiz-Belda, M.-A. (1997). Spontaneous facial behavior during intense emotional episodes: Artistic truth and optical truth. In J. A. Russell & J. M. Fernández-Dols, eds., *The Psychology of Facial Expression*, Cambridge: Cambridge University Press, pp. 255–274.

Fichman, M. (1981) *Alfred Russel Wallace*, Boston: Twayne.

Fieser, J. (1992). Hume's classification of the passions and its precursors. *Hume Studies*, 18(1), 1–17.

Fisher, R. A. (1930). *The Genetical Theory of Natural Selection*, Oxford: Clarendon Press.

Fodor, J. A. (1981). The mind–body problem. *Scientific American*, 244(1), 114–123.

(1983). *The Modularity of Mind*, Cambridge, MA: MIT Press.

Frank, R. (1988). *Passions within Reason: The Strategic Role of the Emotions*, New York: W. W. Norton.

Freeman, R. B., & Gautrey, P. J. (1972). Charles Darwin's *Queries about Expression. Bulletin of the British Museum (Natural History) Historical Series*, 4(3), 205–219. https://doi.org/10.5962/p.310424.

Freeman, R. B. & Gautrey, P. J. (1975). Charles Darwin's *Queries about Expression. Journal of the Society for the Bibliography of Natural History*, 7(3), 259–263.

Frevert, U. (2011). *Emotions in History: Lost and Found*, Budapest: Central European University Press.

Fridlund, A. J. (1992). Darwin's anti-Darwinism in *The Expression of the Emotions in Man and Animals*. In K. T. Strongman, ed., *International Review of Studies of Emotion*, vol. 2, Chichester: Wiley.

(1994). *Human Facial Expression: An Evolutionary View*, London: Academic Press.

(2017). The behavioral ecology view of facial displays, 25 years later. In J.-M. Fernandez-Dols & J. A. Russell, eds., *The Science of Facial Expression*, New York: Oxford University Press, pp. 77–92.

Funkhouser, E. (2017). Beliefs as signals: A new function for belief. *Philosophical Psychology*, 30(6), 809–831.

(2018). Detection, not perception: A reply to Glazer. *Philosophical Psychology*, 31(7), 1120–1125.

Gallagher, S. (2008). Direct perception in the intersubjective context. *Consciousness and Cognition*, 17, 535–543.

Gallagher, S., & Zahavi, D. (2012). *The Phenomenological Mind*, New York: Routledge.

García-Higuera, J.-A., Crivelli, C., & Fernández-Dols, J.-M. (2015). Facial expressions during an extremely intense emotional situation: Toreros' lip funnel. *Social Science Information*, 54(4), 439–454. https://doi.org/10.1177/0539018415596381.

Gendron, M., & Barrett, L. F. (2009). Reconstructing the past: A century of ideas about emotion in psychology. *Emotion Review*, 1(4), 316–39. https://doi.org/10.1177/1754073909338877.

Gert, B. (2010). *Hobbes*, Cambridge: Polity.

Ghiselin, M. T. (1969). *The Triumph of the Darwinian Method*, Berkeley: University of California Press.

Gissis, S. B. (2010). Lamarck on feelings: From worms to humans. In C. T. Wolfe & O. Gal, eds., *The Body as Object and Instrument of Knowledge*, Dordrecht: Springer, pp. 211–239.

Glazer, T. (2017). Looking angry and sounding sad: The perceptual analysis of emotional expression. *Synthese*, 149(9), 3619–3643.

(2018a). The part–whole perception of emotion. *Consciousness and Cognition*, 28, 34–43.
(2018b). Are beliefs signals? *Philosophical Psychology*, 31(7), 1114–1119.
(2019). The social amplification view of facial expression. *Biology and Philosophy*, 34(2), article 33.
(2020). To express or not to express? Ambivalence about emotional expression. In D. Gatzia & G. Brogaard, eds., *The Philosophy and Psychology of Ambivalence: Being of Two Minds*, New York: Routledge, pp. 175–186.
(2021). Emotion regulation and cooperation. *Philosophical Psychology*, 34(8), 1125–1145.
(2023). Expressing 2.0. *Analytic Philosophy*, https://doi.org/10.1111/phib.12308.
(2024). "Trains of thought long associated with action": Charles Darwin's theory of emotion. *History of Philosophy Quarterly*, 41(1), 1–22.
Glover, V., & Hill, J. (2012). Sex differences in the programming effects of prenatal stress on psychopathology and stress responses: An evolutionary perspective. *Physiology & Behavior*, 106(5), 736–740.
Goldie, P. (2000). *The Emotions: A Philosophical Exploration*, Oxford: Oxford University Press.
Goldman, A. (1989). Interpretation psychologized. *Mind & Language*, 4, 161–185.
Gordon, R. M. (1986). Folk psychology as simulation. *Mind & Language*, 1(2), 158–171.
Gould, S. J. (2002). *The Structure of Evolutionary Theory*, Cambridge, MA: Belknap Press of Harvard University Press.
Green, M. S. (2007). *Self-Expression*, New York: Oxford University Press.
(2010a). Perceiving emotions. *Aristotelian Society Supplementary*, 84, 45–61.
(2010b). Replies to Eriksson, Martin and Moore. *Acta Analytica*, 25, 105–117.
(2016). Expressing, showing, and representing. In C. Abell & J. Smith, eds., *Emotional Expression: Philosophical, Psychological, and Legal Perspectives*, New York: Cambridge University Press, pp. 25–45.
(2019). From signaling and expression to conversation and fiction. *Grazer Philosophische Studien*, 96(3), 295–315. doi: https://doi.org/10.1163/18756735-09603002.
Green, O. H. (1970). The expression of emotion. *Mind*, 79(316), 551–568.
Greenway, A. P. (1973). The incorporation of action into associationism: The psychology of Alexander Bain. *Journal of the History of Behavioral Sciences*, 9(1), 42–52.
Griffiths, P. (1997). *What Emotions Really Are: The Problem of Psychological Categories*, Chicago: University of Chicago Press.
Griffiths, P. E., & Scarantino, A. (2005). Emotions in the wild: The situated perspective on emotion. In P. Robbins & M. Aydede, eds., *The Cambridge Handbook of Situated Cognition*, Cambridge: Cambridge University Press, pp. 437–453.

Grinnell, G. (1974). The rise and fall of Darwin's first theory of transmutation. *Journal of the History of Biology*, 7(2), 59–73.
Gruber, H. E., & Barrett, P. H. (1974). *Darwin on Man*, New York: E. P. Dutton.
Haldane, J. B. S. (1932). *The Causes of Evolution*, New York: Longmans, Green.
Hampshire, S. (1976). Feeling and expression In J. Glover, ed., *The Philosophy of Mind*, New York: Oxford University Press, pp. 73–83.
Hanson, N. R. (1958). *Patterns of Discovery: An Inquiry into the Conceptual Foundations of Science*, Cambridge: Cambridge University Press.
Harper, K. (2019). Alexander Bain's *Mind and Body* (1872): An underappreciated contribution to early neuropsychology. *Journal of the History of the Behavioral Sciences*, 55, 139–160.
Harré, R. (1986). An outline of the social constructivist viewpoint. In R. Harré, ed., *The Social Construction of Emotions*, New York: Basil Blackwell, pp. 2–14.
Hartley, D. (1749). *Observations on Man, His Frame, His Duty, and His Expectations*, 2 vols., Bath: Samuel Richardson.
Hartley, L. (2001). *Physiognomy and the Meaning of Expression in Nineteenth-Century Culture*, New York: Cambridge University Press.
Heinroth, O. (1911). Beiträge zur Biologie, namentlich Ethologie und Psychologie der Anatiden. In H. Schalow, ed., *Verhandlungen des 5. Internationalen Ornithologen-Kongresses in Berlin, 30. Mai bis 4. Juni*, Berlin: Deutsche Ornithologische Gesellschaft, pp. 589–702.
Helmholtz, H. (1867). *Handbuch der physiologischen Optik*. In G. Karsten, ed., *Allgemeine Encyklopädie der Physik*, ed. vol. 9. Leipzig: Voss.
Henle, M. (1990). William James and Gestalt psychology. In M. G. Johnson & T. B. Henley, eds., *Reflections on the Principles of Psychology: Williams James after a Century*, Hillsdale, NJ: Erlbaum, pp. 77–99.
Herbert, S. (1974). The place of man in the development of Darwin's theory of transmutation. Part 1. To July 1837. *Journal of the History of Biology*, 7, 217–258.
(1977). The place of man in the development of Darwin's theory of transmutation. Part 2. *Journal of the History of Biology*, 10, 155–227.
Hernández-Avilez, E. G., & Ruiz-Gutiérrez, R. (2023). From one Darwin to another: Charles Darwin's annotations to Erasmus Darwin's 'The Temple of Nature'. *Humanities and Social Sciences Communications*, 10, article 143. https://doi.org/10.1057/s41599-023-01616-y
Hill, C. L. (1951). *A Short History of Modern Philosophy from the Renaissance to Hegel*, Boston: Meador.
Hinde, R. A. (1985). Was "the expression of the emotions" a misleading phrase? *Animal Behaviour*, 33(3), 985–992. https://doi.org/10.1016/S0003-3472(85)80032-4.
Hobbes, T. (1640). *The Elements of Law*. In J. C. A. Gaskin, ed., *The Elements of Law, Natural and Politic*, Oxford: Oxford University Press, 1990.
(2020). *Leviathan*, Norton Critical Edition, 2nd ed., New York: W. W. Norton.

Hodge, M. J. S. (1990). Darwin studies at work: A re-examination of three decisive years (1835–37). In T. H. Levere & W. R. Shea, eds., *Nature, Experiment and the Sciences*, Dordrecht: Kluwer Academic, pp. 249–273.

Hoffman, P. (1991). Three dualist theories of the passions. *Philosophical Topics*, 19(1), 153–200.

Huebner, B., & Glazer, T. (2016). Emotional processing in individual and social recalibration. In J. Kiverstein, ed., *Routledge Companion to the Philosophy of Social Mind*, New York: Routledge, pp. 280–297.

Hughes, S., & Gardner-Thorpe, C. (2022). Charles Bell's (1774–1842) contribution to our understanding of facial expression. *Journal of Medical Biography*, 30(4), 206–214.

Hume, D. (2007). *A Treatise of Human Nature*, vol. 1. D. F. Norton & M. J. Norton, eds., Oxford: Clarendon Press.

Hutto, D. D., Robertson, I., & Kirchhoff, M. D. (2018). A new, better BET: Rescuing and revising basic emotion theory. *Frontiers in Psychology*, 9, article 1217. https://doi.org/10.3389/fpsyg.2018.01217.

Huxley, J. S. (1914). The courtship habits of the Great Crested Grebe. *Proceedings of the Zoological Society of London*, 84(3), 491–562.

(1942). *Evolution: The Modern Synthesis*, London: Allen & Unwin.

Illouz, E. (2013). *Why Love Hurts: A Sociological Explanation*, Cambridge: Polity.

James, S. (1997). *Passion and Action: The Emotions in Seventeenth-Century Philosophy*, New York: Oxford University Press.

James, W. (1884). What is an emotion? *Mind*, 9, 188–205.

(1890). *The Principles of Psychology*, New York: Henry Holt.

Jauss, S. A. (2006). Associationism and taste theory in Archibald Alison's essays. *Journal of Aesthetics and Art Criticism*, 64(4), 415–428.

Johnson, P. (2012). *Darwin: Portrait of a Genius*, New York: Penguin.

Johnson-Laird, P. N., & Oatley, K. (1989) The language of emotions: An analysis of a semantic field, *Cognition and Emotion*, 3(2), 81–123. doi: 10.1080/02699938908408075.

Jordanova, L. J. (1984). *Lamarck*, Oxford: Oxford University Press.

Jørgensen, C. B. (2003). Aspects of the history of the nerves: Bell's theory, the Bell–Magendie law and controversy, and two forgotten works by P. W. Lund and D. F. Eschricht. *Journal of the History of the Neurosciences*, 12(3), 229–249. doi: 10.1076/jhin.12.3.229.16676. PMID: 14628540.

Kallich, M. (1945). The association of ideas and critical theory: Hobbes, Locke, and Addison. *ELH*, 12(4), 290–315.

Keltner, D., Tracy, J. L., Sauter, D., & Cowen, A. (2019). What basic emotion theory really says for the twenty-first century study of emotion. *Journal of Nonverbal Behavior*, 43(2), 195–201. doi: 10.1007/s10919-019-00298-y.

Kennedy, A. (2012). Contextualising the concept of the basic emotion: An historical examination of the theories of Alexander Bain and Herbert Spencer. *History & Philosophy of Psychology*, 14(2), 36–42.

Kenny, A. (1963). *Action, Emotion and Will*, New York: Routledge.

Khalidi, M. A. (2016). Mind-dependent kinds. *Journal of Social Ontology*, 2(2), 223–246.
Kind, A. (2020). *Philosophy of Mind: The Basics*, New York; Routledge.
King, P. (2010). Emotions in medieval thought. In P. Goldie, ed., *The Oxford Handbook of the Philosophy of Emotion*, New York: Oxford, pp. 167–187.
King-Hele, D. (1999). *Erasmus Darwin: A Life of Unequalled Achievement*, London: Giles de la Mare.
(2005). Prologue: Catching up with Erasmus Darwin in the new century. In C. U. M. Smith & R. Arnott, eds., *The Genius of Erasmus Darwin*, Burlington, VT: Ashgate.
Knuuttila, S. (2023). Medieval theories of the emotions. In E. N. Zalta, ed., *The Stanford Encyclopedia of Philosophy* (Fall 2023 edition), https://plato.stanford.edu/archives/fall2023/entries/medieval-emotions/.
Kohn, D. (1980). Theories to work by: Rejected theories, reproduction, and Darwin's path to natural selection. *Studies in the History of Biology*, 4, 67–170.
Koschorke, A. (1999). "Clusters" of ideas: Social interdependence and emotional complexity in David Hartley's *Observations on Man* and Adam Smith's *Theory of Moral Sentiments*. In J. Schlager & G. Stedman, eds., *Representations and Emotions*, Tübingen: Gunther Narr, pp. 113–124.
Koval, P., Kuppens, P., Allen, N. B., & Sheeber, L. (2012). Getting stuck in depression: The roles of rumination and emotional inertia. *Cognition and Emotion*, 26(8), 1412–1427. https://doi.org/10.1080/02699931.2012.667392.
Krause, E. (1879). *Erasmus Darwin. With a Preliminary Notice by Charles Darwin*, London: John Murray.
Krueger, J. (2012). Seeing mind in action. *Phenomenology and the Cognitive Sciences*, 11, 149–173.
(2014). Dewey's rejection of the emotion/expression distinction. In T. Solymosi & J. Shook, eds., *Neuroscience, Neurophilosophy and Pragmatism: Understanding Brains at Work in the World*, New York: Palgrave Macmillan, pp. 140–161.
Krueger, J., & Overgaard, S. (2012). Seeing subjectivity: Defending a perceptual account of other minds. *ProtoSociology*, 47, 239–262.
Kuhn, T. (2000). *The Road Since Structure*, Chicago: University of Chicago Press.
Lakoff, G., & Johnson, M. (1980). *Metaphors We Live By*, Chicago: University of Chicago Press.
Lamarck, J. B. (1984). *Zoological Philosophy: An Exposition with Regard to the Natural History of Animals*, Chicago: University of Chicago Press.
Landes, M. W. (1926). Thomas Brown: Associationist (?).*Philosophical Review*, 35(5), 447–464.
Lennox, J. (2019). Darwinism. In E. N. Zalta, ed., *The Stanford Encyclopedia of Philosophy* (Fall 2019 edition), https://plato.stanford.edu/archives/fall2019/entries/darwinism/.
Leslie, J. C. (2006). Herbert Spencer's contributions to behavior analysis: A retrospective review of *Principles of Psychology*. *Journal of the Experimental Analysis of Behavior*, 86, 123–129.

Lindquist, K. A., Wager, T. D., Kober, H., Bliss-Moreau, E., & Barrett, L. F. (2012). The brain basis of emotion: A meta-analytic review. *Behavioral and Brain Sciences*, 35(3), 121–143.
Loaiza, J. R. (2021). Emotions and the problem of variability. *Review of Philosophy and Psychology*, 12, 329–351. https://doi.org/10.1007/s13164-020-00492-8.
Locke, J. (1894). *An Essay Concerning Human Understanding*, vol. 1, A. C. Fraser, ed., Oxford: Clarendon Press.
Lorch, M., & Hellal, P. (2010). Darwin's "natural science of babies." *Journal of the History of Neuroscience*, 12(2), 140–157. doi: 10.1080/09647040903504823.
Lorde, A. (1984). *Sister Outsider*, Berkeley, CA: Crossing Press.
Lorenz, K. (1939), *Vergleichende Verhaltensforschung. Zoologische Anzeiger*, Supplement, 12, 69–102. doi:10.1007/978-3-7091-3097-1.
(2002). *On Aggression*, New York: Routledge.
Loudon, I. S. L. (1982). Sir Charles Bell and the anatomy of expression. *British Medical Journal (Clinical Research Edition)*, 285(6357), 1794–1796.
Lutz, C. (1988). *Unnatural Emotions: Everyday Sentiments on a Micronesian Atoll and Their Challenge to Western Theory*, Chicago: University of Chicago Press.
Lyons, W. (1980). *Emotion*, New York: Cambridge University Press.
Maas, H. (2003). Where mechanism ends: Thomas Reid on the moral and animal oeconomy. *History of Political Economy*, 35 (Annual Supplement), 338–360.
Mackintosh, J. (1837). *Dissertation on the Progress of Ethical Philosophy*, Edinburgh: Adam and Charles Black.
Madden, M. C., & Madden, E. H. (1978). William James and the problem of relations. *Transactions of the Charles S. Peirce Society*, 14(4), 227–246.
Magendie, F. (1822). Expériences sur les fonctions des racines des nerfs rachidiens. *Journal of Physiology*, 2, 276–279.
Majeed, R. (2023). Does the problem of variability justify Barrett's emotion revolution? *Review of Philosophy and Psychology*, 14, 1421–1441. https://doi.org/10.1007/s13164-022-00650-0.
Manier, E. (1978). *The Young Darwin and His Cultural Circle*, Dordrecht: Reidel.
Maynard Smith, J. (1976). Sexual selection and the handicap principle. *Journal of Theoretical Biology*, 57(1), 239–242.
(1978). The handicap principle: A comment. *Journal of Theoretical Biology*, 70(2), 251–252.
Maynard Smith, J., & Harper, D. (2003). *Animal Signals*, New York: Oxford University Press.
Mayr, E. (1942). *Systematics and the Origin of Species*, New York: Columbia University Press.
(1968). Illiger and the biological species concept. *Journal of the History of Biology*, 1(2), 163–178.
(1994). Typological versus population thinking. In E. Sober, ed., *Conceptual Issues in Evolutionary Biology, 2nd ed., Cambridge, MA: MIT Press, pp. 158–160.*
McGill, V. J., & Welch, L. (1946). A behaviorist analysis of emotions. *Philosophy of Science*, 13(2), 100–122.

McIntyre, J. A. (2000). Hume's passions: Direct and indirect. *Hume Studies*, 26(1), 77–86.
Mead, G. H. (1967). *Mind, Self, & Society: From the Standpoint of a Social Behaviorist*, Chicago: University of Chicago Press.
Merivale, A. (2009). Hume's mature account of the indirect passions. *Hume Studies*, 35(1–2), 185–210.
Merleau-Ponty, M. (2002). *Phenomenology of Perception*, trans. C. Smith, New York: Routledge.
Merton, R. K. (1968) The Matthew effect in science. *Science*, 159(3810), 56–63.
Mill, J. (1829). *Analysis of the Phenomena of the Human Mind*, 2 vols., London: Baldwin and Cradock.
Mill, J. S. (1859). Bain's psychology. *Edinburgh Review*, 110, 287–321.
(1978). *The Collected Works of John Stuart Mill, Volume XI: Essays on Philosophy and the Classics*, J. M. Robson, ed., Toronto: University of Toronto Press; London: Routledge and Kegan Paul.
Millikan, R. (1984). *Language, Thought, and Other Biological Categories*, Cambridge, MA: MIT Press.
Mischel, T. (1966). "Emotion" and "motivation" in the development of English psychology: D. Hartley, James Mill, A. Bain. *Journal of the History of the Behavioral Sciences*, 2(2), 123–144.
Montgomery, W. (1985). Charles Darwin's thought on expressive mechanisms in evolution. In G. Zivin, ed., *The Development of Expressive Behavior: Biology-Environment Interactions*, London: Academic Press, pp. 27–50.
Morell, J. D. (1846). *An Historical and Critical View of the Speculative Philosophy of Europe in the Nineteenth Century*, London: William Pickering.
Morreall, J. (2023). Philosophy of humor. In E. N. Zalta, ed., *Stanford Encyclopedia of Philosophy* (Summer 2023 edition), https://plato.stanford.edu/archives/sum2023/entries/humor/.
Morton, A. (1980). *Frames of Mind: Constraints on the Commonsense Conception of the Mental*, Oxford: Clarendon Press.
Na'aman, O. (2020). The fitting resolution of anger. *Philosophical Studies*, 177(8), 2417–2430.
(2021). The rationality of emotional change: Toward a process view. *Nous*, 55(2), 245–269.
Naar, H. (2024). Are emotions events, processes, mechanisms or dispositions? In A. Scarantino, ed., *Emotion Theory: The Routledge Comprehensive Guide, vol. 1: History, Contemporary Theories, and Key Elements*, New York: Routledge.
Nanay, B. (2010). Population thinking as trope nominalism. *Synthese*, 177, 91–109.
Neher, A. (2008). Sir Charles Bell and the anatomy of expression, *RACAR: Revue d'art canadienne/Canadian Art Review*, 33(1–2), 59–65. doi:10.7202/1069547ar.
Newen, A., Welpinghus, A., & Juckel, G. (2015). Emotion recognition as pattern recognition: The relevance of perception. *Mind & Language*, 30(2), 113–233.

Newton I. (1803). *The Mathematical Principles of Natural Philosophy*, vol. 2, A. Motte, trans., London: Symonds.

Ng, B. C., Cui, C., & Cavallaro, F. (2019) The annotated lexicon of Chinese emotion words, *Word*, 65(2), 73–92, doi: 10.1080/00437956.2019.1599543.

Nolen-Hoeksema, S. (2000). The role of rumination in depressive disorder and mixed anxiety/depressive symptoms. *Journal of Abnormal Psychology*, 109, 504–511.

Nussbaum, M. C. (2001). *Upheavals of Thought: The Intelligence of Emotions*, Cambridge: Cambridge University Press.

(2016). *Anger and Forgiveness*, Oxford: Oxford University Press.

Oberg, B. B. (1976). David Hartley and the association of ideas. *Journal of the History of Ideas*, 37, 441–454.

Odenbaugh, J. (n.d.) In a sentimental mood. Unpublished manuscript.

O'Grady, J. (2005). Review of From Passions to Emotions: The Creation of a Secular Psychological Category, by Thomas Dixon, and *The Navigation of Feeling: A Framework for the History of Emotions*, by William M. Reddy. *Philosophy*, 80(1), 156–159.

Orlandi, N. (2011). The innocent eye: Seeing-as without concepts. *American Philosophical Quarterly*, 48(1), 17–31.

Ortony, A., & Turner, T. J. (1990). What's basic about basic emotions? *Psychological Review*, 97(3), 315–331. https://doi.org/10.1037/0033-295X.97.3.315.

Osborn, H. F. (1894). *From the Greeks to Darwin: An Outline of the Development of the Evolution Idea*, New York: Macmillan.

Ospat, D. (1981). *The Development of Darwin's Theory: Natural History, Natural Theology, and Natural Selection, 1838–1859*, Cambridge: Cambridge University Press.

Overgaard, S. (2007). *Wittgenstein and Other Minds: Rethinking Subjectivity and Intersubjectivity with Wittgenstein, Levinas, and Husserl*, New York: Routledge.

Panksepp, J. (1992). A critical role for "affective neuroscience" in resolving what is basic about basic emotions. *Psychological Review*, 99(3), 554–560. https://doi.org/10.1037/0033-295X.99.3.554.

(2008). Carving "natural" emotions: "Kindly" from bottom-up but not top-down. *Journal of Theoretical and Philosophical Psychology*, 28(2), 395–422.

(2011). The basic emotional circuits of mammalian brains: Do animals have affective lives? *Neuroscience and Biobehavioral Reviews*, 35(9), 1791–1804.

Panksepp, J., & Burgdorf, J. (2003). "Laughing" rats and the evolutionary antecedents of human joy? *Physiology & Behavior*, 79(3), 533–547. doi: 10.1016/s0031-9384(03)00159-8.

Panksepp, J., & Watt, D. (2011). What is basic about basic emotions? Lasting lessons from affective neuroscience. *Emotion Review*, 3(4), 387–396.

Penn, D. J., & Számadó, S. (2020). The Handicap Principle: How an erroneous hypothesis became a scientific principle. *Biological Reviews of the Cambridge Philosophical Society*, 95(1), 267–290.

Pettigrove, G. (2012). Meekness and "moral" anger. *Ethics*, 122, 341–370.
Pitcher, G. (1965). Emotion. *Mind*, 74(295), 326–346.
Plamper, J. (2017). *The History of Emotions: An Introduction*, New York: Oxford University Press.
Plutchik, R. (1962). *The Emotions: Facts, Theories and a New Model*, Crown Publishing Group/Random House.
(2001). The nature of emotions. *American Scientist*, 89(4), 344–350.
Porter, D. (2018a) *Science, Form, and the Problem of Induction in British Romanticism*, Cambridge: Cambridge University Press.
(2018b) Epistemic images and vital nature: Darwin's Botanic Garden as image text book. *European Romantic Review*, 29(3), 295–308. doi: 10.1080/10509585.2018.1465717.
Pott, H. J. (2024). James's revolutionary theory of emotions. In A. M. Klein, ed., *The Oxford Handbook of William James*, online edition, Oxford: Oxford University Press, pp. 41–60.
Priestley, J. (1775). *Hartley's Theory of the Human Mind, on the Principle of the Association of Ideas; with Essays Relating to the Subject of It*, London: J. Johnson.
(1782). *Institutes of Natural and Revealed Religion*, London: Joseph Johnson.
Prinz, J. J. (2004). *Gut Reactions: A Perceptual Theory of Emotion*, Oxford: Oxford University Press.
(2012). Emotions: How many are there? In E. Margolis, R. Samuels, & S. P. Stitch, eds., *The Oxford Handbook of the Philosophy of Cognitive Science*, New York: Oxford University Press, pp. 183–200.
Prodger, P. (2009). *Darwin's Camera: Art and Photograph in the Theory of Evolution*, New York: Oxford University Press.
Radick, G. (2010). Darwin's puzzling *Expression*. *Comptes Rendus Biologies*, 333(2), 181–187. https://doi.org/10.1016/j.crvi.2009.12.006.
Rapaport, D. (1974). *The History of the Concept of Association of Ideas*, New York: International Universities Press.
Ratner, C. (1989). A social constructionist critique of the naturalistic theory of emotion. *Journal of Mind and Behavior*, 10(3), 211–239.
Reddy, W. (2001). *The Navigation of Feeling: A Framework for the History of Emotions*, Cambridge: Cambridge University Press.
Reid, T. (1827). *Essays on the Powers of the Human Mind; To Which Are Added, an Essay on Quantity, and an Analysis of Aristotle's Logic*, London: Thomas Tegg.
Reisenzein, R. (1995). On Oatley and Johnson–Laird's theory of emotion and hierarchical structures in the affective lexicon, *Cognition and Emotion*, 9(4), 383–416. doi: 10.1080/02699939508408973.
Ribot, T.-A. (1874). *English Psychology – Hartley – James Mill – Herbert Spencer – A. Bain – G. H. Lewes – Samuel Bailey – John Stuart Mill*, New York; D. Appleton.
Richards, R. J. (1992). *The Meaning of Evolution: The Morphological Construction and Ideological Reconstruction of Darwin's Theory*, Chicago: University of Chicago Press.

Richards, R. (2009). Darwin on mind, morals and emotions. In J. Hodge & G. Radick, eds., *The Cambridge Companion to Darwin*, New York: Cambridge University Press, pp. 96–119.

(2005) Darwin's metaphysics of mind. In V. Hoesle & C. Illies, eds., *Darwin and Philosophy*, South Bend, IN: Notre Dame University Press, pp. 166–180.

Rieber, R. W. (2012). Psychology of Alexander Bain. In R. W. Rieber, ed., *Encyclopedia of the History of Psychological Theories*, New York: Springer.

Robinson, J. (2018). Emotion as process. In H. Naar and F. Teroni, eds., *The Ontology of Emotions*, Cambridge: Cambridge University Press, pp. 51–70.

Rock, I. (1983). *The Logic of Perception*, Cambridge, MA: MIT Press.

Rogers, J. A. (1972). Darwinism and social Darwinism. *Journal of the History of Ideas*, 33(2), 265–280.

Rosenwein, B. H. (2006). *Emotional Communities in the Early Middle Ages*, Ithaca, NY: Cornell University Press.

(2015). *Generations of Feeling: A History of Emotions, 600–1700*, Cambridge: Cambridge University Press.

(2020). *Anger: The Conflicted History of an Emotion*, New Haven, CT: Yale University Press.

Rowson, R. (forthcoming). Perceiving the event of emotion. *Ergo*.

Russell, B. (1946). *A History of Western Philosophy*, London: George Allen and Unwin.

Russell, J. A. (1991). In defense of a prototype approach to emotion concepts. *Journal of Personality and Social Psychology*, 60(1), 37–47. https://doi.org/10.1037/0022-3514.60.1.37.

(2003). Core affect and the psychological construction of emotion. *Psychological Review*, 110(1), 145–72. https://doi.org/10.1037/0033-295x.110.1.145.

Russell, J. A., & Fehr, B. (1994). Fuzzy concepts in a fuzzy hierarchy: Varieties of anger. *Journal of Personality and Social Psychology*, 67(2), 186–205. https://doi.org/10.1037/0022-3514.67.2.186.

Russell, J. A., & Fernández-Dols, J. M. (1997). What does a facial expression mean? In J. A. Russell & J. M. Fernández-Dols, eds., *The Psychology of Facial Expression*, Cambridge: Cambridge University Press, pp. 3–30.

Ryle, G. (2009). *The Concept of Mind*, New York: Routledge.

Sartre, J. P. (1948). *The Emotions: Outline of a Theory*, B. Frechtman, trans., New York: Philosophical Library.

Scarantino, A. (2014). The motivational theory of emotion. In J. D'Arms & D. Jacobson, eds., *Moral Psychology and Human Agency*, Oxford: Oxford University Press, pp. 156–185.

(2015). Basic emotions, psychological construction, and the problem of variability. In L. F. Barrett and J. A. Russell, eds., *The Psychological Construction of Emotion*, New York: Guilford Press, pp. 334–376.

(2017). How to do things with emotional expressions: The theory of affective pragmatics. *Psychological Inquiry*, 28(2–3), 165–185.

(2019). Affective pragmatics extended: From natural to overt expressions of emotion. In U. Hess & S. Hareli, eds., *The Social Nature of Emotion*

Expression: What Emotions Can Tell Us about the World, Cham: Springer Nature, pp. 49–82.
Scarantino, A., & de Sousa, R. (2018). Emotion. In E. N. Zalta, ed., *Stanford Encyclopedia of Philosophy* (Summer 2021 edition), https://plato.stanford.edu/archives/sum2021/entries/emotion/.
Scarantino, A., & Griffiths, P. (2011). Don't give up on basic emotions. *Emotion Review*, 3(4), 444–454. https://doi.org/10.1177/1754073911410745.
Schachter, S. & Singer, J. (1962). Cognitive, social and physiological determinants of emotional states. *Psychological Review*, 69(5), 379–399. doi:10.1037/h0046234.
Scheler, M. (1954). *The Nature of Sympathy*, P. Heath, trans., London: Routledge & Kegan Press.
Schmitter, A. M. (2021a). Hobbes on the emotions. Supplement to: 17th and 18th century theories of emotions. In E. N. Zalta, ed., *The Stanford Encyclopedia of Philosophy* (Summer 2021 edition), https://plato.stanford.edu/entries/emotions-17th18th/LD3Hobbes.html.
(2021b). Hume on the emotions. Supplement to: 17th and 18th century theories of emotions. In E. N. Zalta, ed., *The Stanford Encyclopedia of Philosophy* (Summer 2021 edition), https://plato.stanford.edu/entries/emotions-17th18th/LD8Hume.html.
Schofield, R. E. (1963). *The Lunar Society of Birmingham: A Social History of Provincial Science and Industry in Eighteenth-Century England*, New York: Oxford University Press.
Schwartz, S. P. (2012). *A Brief History of Analytic Philosophy: From Russell to Rawls*, Hoboken, NJ: Wiley-Blackwell.
Schweber, S. S. (1977). The origin of the *Origin* revisited. *Journal of the History of Biology*, 10, 229–316.
Shapin, S. (1975). Phrenological knowledge and the social structure of early nineteenth century Edinburgh. *Annals of Science*, 32, 245–256.
Shargel, D., & Prinz, J. (2017). An enactivist theory of emotional content. In H. Naar & F. Teroni, eds., *The Ontology of Emotions*, Cambridge: Cambridge University Press, pp. 110–129.
Simili, R. (2005). Two special doctors: Erasmus Darwin and Luigi Galvani. In C. U. M. Smith & R. Arnott, eds., *The Genius of Erasmus Darwin*, Burlington, VT: Ashgate, pp. 145–160.
Simpson, G. G. (1949). *The Meaning of Evolution: A Study of the History of Life and of Its Significance for Man*, New Haven, CT: Yale University Press.
Slater, M. (2015). Natural kindness. *British Journal for the Philosophy of Science*, 66, 375–411.
Smith, C. U. M. (1987). David Hartley's Newtonian neuropsychology. *Journal for the History of Behavioral Science*, 23, 123–136.
(2005). All from fibers: Erasmus Darwin's evolutionary psychobiology. In C. U. M. Smith & R. Arnott, eds., *The Genius of Erasmus Darwin*, Burlington, VT: Ashgate, pp. 133–144.

(2010). Like grandfather, like grandson: Erasmus and Charles Darwin on evolution. *Perspectives in Biology and Medicine*, 53(2), 186–199.
Smith, S. (1960). The origin of the *Origin* as discerned from Charles Darwin's notebooks and his annotations in the books he read between 1837 and 1842. *Advancement of Science*, 16, 391–401.
Soames, S. (2003a). *Philosophical Analysis in the Twentieth Century, vol. 1: The Dawn of Analysis*, Princeton, NJ: Princeton University Press.
(2003b). *Philosophical Analysis in the Twentieth Century, vol. 1: The Age of Meaning*, Princeton, NJ: Princeton University Press.
Sober, E. (1980). Evolution, population thinking, and essentialism. *Philosophy of Science*, 47(3), 350–383.
Sodano, J. P. (2017). Uneasy passions. *Eighteenth Century*, 58(4), 449–467.
Solomon, R. C. (1977). The logic of emotion. *Nous*, 11(1), 41–49.
(2008). The philosophy of emotions. In M. Lewis, J. M. Haviland-Jones, & L. F. Barrett, eds., *Handbook of Emotions*, New York: Guilford Press, pp. 3–16.
Spencer, H. (1855). *Principles of Psychology*, London: Longman, Brown, Green, and Longmans.
(1860). The physiology of laughter. *Macmillan's Magazine,* March 1860, 395–402.
Srinivasan, A. (2018). The aptness of anger. *Journal of Political Philosophy*, 26(2), 123–144.
Stamos, D. N. (2007). *Darwin and the Nature of Species*, Albany: SUNY Press.
Stauffer, R. C. (1975). *Charles Darwin's Natural Selection, Being the Second Part of His Big Species Book Written from 1856 to 1858*, Cambridge: Cambridge University Press.
Stearns, C. Z., & Stearns, P. N. (1986). *Anger: The Struggle for Emotional Control in America's History*, Chicago: University of Chicago Press.
Stearns, P. (1993). Boys, girls, and emotions: Redefinitions and historical change. *Journal of American History*, 80(1), 36–74.
(1994). *American Cool: Constructing a Twentieth-Century Emotional Style*, New York: NYU Press.
Stearns, P. N., & Stearns, C. Z. (1985). Emotionology: Clarifying the history of emotion and emotional standards. *American Historical Review*, 90(4), 813–836.
Stebbins, G. L. (1966). *Processes of Organic Evolution*, Englewood Cliffs, NJ: Prentice Hall.
Steward. H. (1997). *The Ontology of Mind: Events, Processes, and States*, Oxford: Oxford University Press.
Stout, R. (2010). Seeing the anger in someone's face. *Aristotelian Society Supplementary*, 84, 29–43.
Strack, F., Martin, L. L., & Stepper, S. (1988). Inhibiting and facilitating conditions of the human smile: A nonobtrusive test of the facial feedback hypothesis. *Journal of Personality and Social Psychology*, 54(5), 768–777. https://doi.org/10.1037/0022-3514.54.5.768.

Strawson, P. (1962). Freedom and resentment. *Proceedings of the British Academy*, 48, 187–211.

Swisher, C. N. (1967). Charles Darwin on the origins of behavior. *Bulletin of the History of Medicine*, 41(1), 24–43.

Tabb, K. (2019). Locke on enthusiasm and the association of ideas. *Oxford Studies in Early Modern Philosophy*, 9, 75–104.

Taylor, C. (1979). Action as expression. In C. Diamond & J. Teichmann, eds., *Intention and Intentionality: Essays in Honour of G. E. M. Anscombe*, Brighton: Harvester Books, pp. 73–89.

Thébert, A. (2014). David Hartley: Vibrations, associations, actions. *Dix-huitième siècle*, 46(1), 417–438. doi:10.3917/dhs.046.0417.

Tinbergen, N. (1963). On aims and methods of ethology. *Zeitschrift für Tierpsychologie*, 20(4), 410–433.

Titchener, E. B. (1928). *A Text-Book of Psychology*, New York: Macmillan.

Tobin, E. (2017). Mechanisms and natural kinds. In S. Glennan & P. Illari, eds., *The Routledge Handbook of Mechanisms and Mechanical Philosophy*, Abingdon: Taylor and Francis.

Tomkins, S. (1962). *Affect, Imagery, Consciousness, vol. 1: The Positive Affects*, New York: Springer.

(1963). *Affect, Imagery, Consciousness, vol. 2: The Negative Affects*, New York: Springer.

Tormey, A. (1971). *The Concept of Expression*, Princeton, NJ: Princeton University Press.

Townsend, D. (1993). The aesthetics of Joseph Priestley. *Journal of Aesthetics and Art Criticism*, 51(4), 561–571. https://doi.org/10.2307/431889.

Tuck, R. (1996). Hobbes's moral philosophy. In T. Sorell, ed., *The Cambridge Companion to Hobbes*, Cambridge: Cambridge University Press, pp. 175–207. doi:10.1017/CCOL0521410193.009.

Turteltaub, J. (2007). National Treasure: Book of Secretes [film], Walt Disney Studios.

Uglow, J. (2005). But what about the women? The Lunar Society's attitude to women and science, and to the education of girls. In C. U. M. Smith & R. Arnott, eds., *The Genius of Erasmus Darwin*, Burlington, VT: Ashgate, pp. 163–178.

Valsania, M. (2005). "Another and the same": Nature and human beings in Erasmus Darwin's doctrines of love and imagination. In C. U. M. Smith & R. Arnott, eds., *The Genius of Erasmus Darwin*, Burlington, VT: Ashgate, pp. 337–356.

van Gijn, J. (2011). Charles Bell (1774–1842). *Journal of Neurology*, 258, 1189–1190. https://doi.org/10.1007/s00415-011-5912-5.

Van Kleef, G. A. (2009). How emotions regulate social life: The emotions as social information (EASI) model. *Current Directions in Psychological Science*, 18(3), 184–188.

Viola, M. (2022). Seeing through the shades of situated affectivity: Sunglasses as a socio-affective artifact. *Philosophical Psychology*, 1–25. https://doi.org/10.1080/09515089.2022.2118574.

Wade, N. J. (2005). The persisting vision of David Hartley (1705–1757). *Perception*, 34(1), 1–6. doi: 10.1068/p3401ed.

Wallace, A. R. (1867). Letter no. 5437. *Darwin Correspondence Projects*, www.darwinproject.ac.uk/letter/?docId=letters/DCP-LETT-5437.xml (accessed September 13, 2023).

(1873). Review: *The Expression of the Emotions in Man and Animals*. *Quarterly Journal of Science*, 3(37), 113–118.

(1905). *My Life: A Record of Events and Opinions*, vol. 2, London: Chapman and Hall.

Walls, J. (1982). The psychology of David Hartley and the root metaphor of mechanism: A study in the history of psychology. *Journal of Mind and Behavior*, 3(3), 259–274

Walsh, R. T. G. (2017). David Hartley's enlightenment psychology: From association to sympathy, theopathy, and moral sensibility. *Journal of Theoretical and Philosophical Psychology*, 37(1), 48–63.

Wang, Y., Liu, H., & Sun, Z. (2017). Lamarck rises from his grave: Parental environment-induced epigenetic inheritance in model organisms and humans. *Biological Reviews*, 92(4), 2084–2111.

Warren, H. C. (1921). *A History of the Association Psychology*, New York: Charles Scribner's Sons.

Wells, H. G., Huxley, J. S., & Wells, G. P. (1934). *The Science of Life*, New York: Literary Guild.

Wierzbicka, A. (1992) Talking about emotions: Semantics, culture, and cognition. *Cognition and Emotion*, 6(3–4), 285–319. doi: 10.1080/02699939208411073.

Winther, R. G. (2000). Darwin on variation and heredity. *Journal of the History of Biology*, 33(3), 425–455.

Wittgenstein, L. (1980). *Remarks on the Philosophy of Psychology*, Oxford: Blackwell.

(2006). *Philosophical Investigations*, Malden, MA: Blackwell.

Woody, W. D. (1999). William James and Gestalt psychology. *Journal of Mind and Behavior*, 20(1), 79–92.

Workman, L., & Reader, W. (2014). *Evolutionary Psychology*, 3rd ed., New York: Cambridge University Press.

Young, R. M. (1970). *Mind, Brain, and Adaptation in the Nineteenth Century*, New York: Oxford University Press.

Zahavi, A. (1975). Mate selection: A selection for handicap. *Journal of Theoretical Biology*, 53(1), 205–214.

(1977). The cost of honesty. *Journal of Theoretical Biology*, 67(3), 603–605.

Zahavi, A., & Zahavi, A. (1997). *The Handicap Principle: A Missing Piece of Darwin's Puzzle*, New York: Oxford University Press.

Zirkle, C. (1935). The inheritance of acquired characters and the provisional hypothesis of pangenesis. *American Naturalist*, 69(724), 417–445.

(1946). The early history of the idea of the inheritance of acquired characters and of pangenesis. *Transactions of the American Philosophical Society*, 35(2), 91–151.

Index

For EU product safety concerns, contact us at Calle de José Abascal, 56–1°,
28003 Madrid, Spain or eugpsr@cambridge.org.

www.ingramcontent.com/pod-product-compliance
Ingram Content Group UK Ltd.
Pitfield, Milton Keynes, MK11 3LW, UK
UKHW021527130126
466887UK00008B/111

* 9 7 8 1 0 0 9 5 3 8 2 4 4 *